SpringerBriefs in Earth Sciences

Series Editor

Pedro Maciel de Paula Garcia, Universidade Federal de Mato Grosso, Cuiabá, Mato Grosso, Brazil

SpringerBriefs in Earth Sciences present concise summaries of cutting-edge research and practical applications in all research areas across earth sciences. It publishes peer-reviewed monographs under the editorial supervision of an international advisory board with the aim to publish 8 to 12 weeks after acceptance. Featuring compact volumes of 50 to 125 pages (approx. 20,000–70,000 words), the series covers a range of content from professional to academic such as:

- timely reports of state-of-the art analytical techniques
- bridges between new research results
- snapshots of hot and/or emerging topics
- literature reviews
- in-depth case studies

Briefs will be published as part of Springer's eBook collection, with millions of users worldwide. In addition, Briefs will be available for individual print and electronic purchase. Briefs are characterized by fast, global electronic dissemination, standard publishing contracts, easy-to-use manuscript preparation and formatting guidelines, and expedited production schedules.

Both solicited and unsolicited manuscripts are considered for publication in this series.

Joseph Davidovits

Ancient Geopolymers in South America and Easter Island

 Springer

Joseph Davidovits
Geopolymer Institute–Institut Géopolymère
Saint-Quentin, France

ISSN 2191-5369   ISSN 2191-5377   (electronic)
SpringerBriefs in Earth Sciences
ISBN 978-3-031-75335-0   ISBN 978-3-031-75336-7   (eBook)
https://doi.org/10.1007/978-3-031-75336-7

Photo Credits: All photos by Ralph Davidovits (RD) unless otherwise noted.
Graphics Credits: All graphics by the author.

This Springer imprint is published by the registered company Springer Nature Switzerland AG
The registered company address is: Gewerbestrasse 11, 6330 Cham, Switzerland

If disposing of this product, please recycle the paper.

Acknowledgments

First, my love and appreciation to my sons Ralph Davidovits and Frederic Davidovits, who work with me at the Geopolymer Institute. Ralph has travelled to the archaeological sites and taken most of the photographs; Frederic has searched the relevant archaeological and scientific literature. My thanks to Carolina Frêne, a Chilean freelance journalist, who edited my French manuscript before translating it into Latin American Spanish. I would also like to acknowledge the role of my friend Andrew Claude James, a retired lecturer in Technical English at our local technical university, INSSET, who has been editing my English scientific manuscripts since 1981.

Contents

Chapter 1
Why Did We Have to Wait 40 Years?

Why did we have to wait 40 years?—The 1979 Congress of Egyptologists—My Meeting with Francisco Aliaga—The New York Congress in 1981—Back home, the research continues—The Congress in Bradford, England, U.K. in 1982—Waking up in 2016: Ralph's visit to Easter Island—Early travelers describe statues made of artificial stone.

It is Thursday, February 22, 2018, and I am in the geopolymer materials analytical laboratory of Pyromeral Systems SA, just outside of Paris. I am standing in front of a scanning electron microscope (SEM) screen, ready to dive into the analysis of the sample we refer to as PP1-A. This particular sample is a piece of a supposed volcanic rock from the Bolivian archaeological site, Pumapunku (refer to the map in Fig. 2. 3). Situated near Tiahuanaco/Tiwanaku and its renowned Puerta del Sol (Gate of the Sun), Pumapunku is steeped in history and mystery.

Today, we focus on conducting a detailed study of this intriguing sample. PP1-A was extracted from the remnants of a remarkable H-shaped structure, believed to have been constructed using alien technology, according to local oral tradition and our knowledgeable guides. Using an optical microscope, I had personally examined this sample before and stumbled across a peculiar anomaly that piqued my curiosity. Now, Mathilde, one of our skilled laboratory professionals in charge of the electron microscope, has aligned the electron beam precisely on point No. 4 of the sample.

As I gaze at the screen, the image reveals a wealth of information. The crystals of plagioclase, hornblende, and pyroxene, minerals typically found in this type of volcanic rock, are exquisitely detailed. However, there is something more captivating. Situated within the crystals, there is a peculiar grayish mass resembling gum. I indicate the exact spot to Mathilde, and she skillfully adjusts the beam to explore further.

The scanning electron microscope offers us two distinct advantages. Firstly, the image is incredibly sharp, almost giving the illusion of depth. Additionally, with the

J. Davidovits, *Ancient Geopolymers in South America and Easter Island*, SpringerBriefs in Earth Sciences, https://doi.org/10.1007/978-3-031-75336-7_1

help of EDS (Energy Dispersive X-ray Spectroscopy) or WDS (Wavelength Dispersive Spectroscopy) analyzers and specialized software, we can estimate the elementary composition of a given sample. Here, we carried out EDS analyses on our sample PP1-A. It typically takes around 10 s, during which a spectrum graph appears, showcasing peaks and curves attributed to each atom: carbon, oxygen, sodium, aluminum, silicon, iron, and so on. Suddenly, Mathilde draws my attention to something unexpected. With her pencil pointing to the screen, she highlights a low-intensity peak between the carbon and oxygen peaks. "What is this?" she asks insistently.

With a smile on my face, I watched without a word before finally breaking the silence. "It's nitrogen (N), Mathilde. We have discovered something extraordinary," I announced. The presence of carbon (C) and nitrogen (N) in our presumably volcanic rock sample is concrete evidence of the existence of organic matter. Under normal circumstances, this would be deemed impossible. Therefore, it is reasonable to conclude that this "andesite" from Pumapunku is an artificial rocky material crafted by human hands.

The significance of this discovery cannot be overstated. It brings closure to my investigations, which began four decades ago but were abandoned due to the lack of positive results at the time. Back then, my scientific pursuits revolved around the science of mineral geosynthesis, which later found its definitive name with the advent of Geopolymers.

However, this was not an easy task because my scientific background is rooted in a totally different field. In fact, after graduating with a degree in Chemical Engineering from the Ecole Nationale Supérieure de Chimie in Rennes, France, in 1958 and a Doctorate in Organic Polymer Chemistry from the University of Mainz, Germany, in 1960, I worked in research on organic polymers for the textile and leather industries in France between 1962 and 1972. During this time, I received an annual prize from the French Society of Textile Chemistry in 1964 for my work on synthetic textile fibers. In 1969, I was recruited by the French textile company Delcer Industries to set up a research laboratory. Nevertheless, following several catastrophic fires involving common organic plastics and synthetic textiles between 1970 and 1972, I decided to focus my research on new heat, nonflammable, and noncombustible plastic materials. I set up the private research company Cordi (Coordination and Development of Innovation) in 1972 in Saint-Quentin, Picardie. In 1979, I published my discovery of the geopolymer concept and created the nonprofit Geopolymer Institute (in French: Institut Géopolymère).

Geopolymers are synthetic mineral substances produced by chemical processes that are actually also found in nature. Yet, these man-made stones are essentially re-agglomerated materials. The principle is simple: starting with a mineral substance such as eroded, disintegrated, or naturally fragmented rock, we employ a binder—a geological glue, if you will—to compactly bind the mineral particles together. The final result is a rock that appears entirely natural. Even a geologist would fail to detect anything unusual. It is only through close examination of the binder that the synthetic nature of the rock becomes apparent, as the particles themselves are unmistakably limestone, granite, sand, or whichever material we choose to work with.

In June 1974, my scientific partners and I made an intriguing discovery at the Mineralogy Laboratory of the Museum of Natural History in Paris. We had just created a unique sandstone with a high concentration of quartz. One day, I jokingly asked my colleagues what would happen if we buried a piece of our synthesized product in the garden of our laboratory and an archaeologist uncovered it 3000 years later. Their response left me astounded: the analysis would indicate that this object, unearthed from the ruins of Saint-Quentin (our hometown), originated from a natural outcrop in southern Egypt! On that day, I realized that if I did not disclose the synthetic nature of our creation, it would be mistaken for a natural stone and classified as such.

This revelation sparked a thought in my mind. Could the ancient megaliths, those massive stone structures of antiquity, also be made of artificial stone? Why were the quarries situated many kilometers away from the archaeological sites? Why were we speculating about the intricate mechanics of transportation and the sheer size of these colossal stone blocks? The notion that they might have been "manufactured" did not seem so far-fetched. Admittedly, given the superficial nature of the megalith analysis conducted thus far, it is evident that no one has yet questioned their presumed natural characteristics.

On this topic, I vividly recall being deeply impressed by the book "Fantastic Easter Island" (Fantastique Ile de Pâques) written by the French ethnologist Francis Mazière during those years (originally published in 1965). His assertions left me pondering the construction techniques employed for the island's statues. Motivated and intrigued by his writings, I reached out to F. Mazière in early 1974, and we had several meetings at the Robert Laffont publishing house in Paris to discuss the possibilities of this intriguing "manufacturing technique." Unfortunately, besides the initial conversations that held promising clues, there was no follow-up to delve further into these discussions.

However, my interest in continuing my research in this field remained unwavering. I decided to contact UNESCO and met with the head of the Service for the Protection of World Monuments and Sites. Through this encounter, I obtained a copy of the report issued by the UNESCO expedition that had visited Easter Island in 1972. This report contained detailed mineralogical analyses of the island's iconic statues, the famous Moai. I firmly believed this study could provide scientific evidence supporting the theory that these statues were created through agglomeration. At that time, my research focused specifically on the oldest statues, those from the first and second periods, excluding the more recent ones found in the quarries, as their structure clearly matched that of rocks originating from the volcano. Although the clues gathered were sufficient to continue my investigations, I recognized the need for more comprehensive information to progress further in my research.

At the end of 1974, I found myself fully immersed in managing and furthering my two greatest passions: the industrial applications of Geopolymers and their potential to shed light on archaeological enigmas and mysteries. As time passed, I dedicated myself to exploring these fields further, and in September 1979, I established the Geopolymer Institute (Institut Géopolymère) in my hometown of Saint-Quentin, France. To this day, it remains the official platform for my scientific endeavors in these areas.

1.1   The 1979 Congress of Egyptologists

In September 1979, an exciting event took place: the Second International Congress of Egyptologists held in Grenoble, France. During this congress, I had the opportunity to present my groundbreaking theory on the construction of the pyramids using agglomerated stone. Instead of the conventional belief that the pyramids were carved, transported using sleds, and raised with immense ramps by an army of enslaved people, I proposed that they were actually created on-site using molds, much like the process of working with concrete.

To my surprise, the reaction from Egyptologists was not openly hostile. While they acknowledged that the ancient Egyptians possessed the knowledge to craft intricate stone vases, enamels, and other small objects, they adamantly refused to accept that this knowledge could be applied to the grand-scale construction of monumental works. This stance has remained unchanged over the years despite the growing body of evidence supporting my theory.

Since that momentous congress 40 years ago, numerous colleagues, scientists, and researchers worldwide have published scientific studies that validate the theory I put forth all those years ago. The accumulation of evidence continues to strengthen the case for a new understanding of how these magnificent structures were constructed.

1.2   My Meeting with Francisco Aliaga

In early 1980, numerous French newspapers dedicated articles to my groundbreaking work. These publications caught the attention of Francisco Aliaga, a Peruvian anthropologist who reached out to me in July of that same year. He expressed his eagerness to arrange a meeting and discuss our shared interests. Eventually, we convened at the headquarters of my laboratory in Saint-Quentin.

Francisco Aliaga, hailing from a region in central Peru between the Peruvian capital, Lima, and the captivating city of Cuzco, had come to France as part of a university exchange program with the University of Orleans, located south of Paris. During our meeting, Francisco revealed his admiration for my theory on agglomerated stone in the Egyptian pyramids. I vividly recall his resolute declaration as we explored the premises of the Geopolymer Institute, "Joseph, you are absolutely correct!".

To support my theory, Francisco recounted his encounters with ancient Peruvian craftsmen during his anthropological work at the National Institute of Culture in Huankayo. These artisans possessed a remarkable ability to create intricate stone objects using a traditional technique that involved softening the stone through the use of plant extracts. With great excitement, he took a small black sculpture from his pocket, exemplifying the method he described. According to Francisco, the indigenous descendants of the Incas passed down stories of the builders of the magnificent megalithic fortresses of Sacsayhuaman, near Cuzco, employing a similar technique to render the rock malleable with plant extracts.

I listened intently to Francisco's account, captivated by its striking resemblance to the chemical reactions we had been replicating in our laboratory research on Geopolymers. Remarkably, it seemed that multiple civilizations had utilized this approach, even in geographically distant locations far from the Altiplano and the region surrounding Lake Titicaca. As I go further into my narrative, I will shed light on the fascinating connection between this chemical science and civilizations spanning the globe.

1.3 The New York Congress in 1981

Given our shared passion and curiosity, I proposed to Francisco that we embark on a scientific collaboration leading up to the Archaeometry 21 International Congress. Taking place in Brookhaven, near New York, by the following year, this conference presented an ideal platform for us to share our research findings and engage with fellow experts in the field. It was July 1980, and to secure our participation, it was imperative to register by October 1., 1980.

We were both excited about the opportunity to share our work and theories in our field with colleagues from around the globe. With that in mind, I wasted no time and promptly submitted our formal registration to confirm our participation. I personally registered an initial presentation on ancient pottery research, which I collaborated on with the French scientist Liliane Courtois from the Institute of Archaeological Research in Paris.

From that point on, our schedules became packed as we diligently prepared our papers for the upcoming congress. The first presentation, a collaborative effort with Liliane Courtois, was in the form of a poster marked as number 16 in the official program (see Fig. 1.1). It was scheduled for Monday, May 18, in the afternoon. The title of our presentation was "Differential Thermal Analysis (DTA) Detection of Intra-Ceramic Geopolymeric Setting in Archaeological Ceramics and Mortars."

The following day, on May 19, I had the privilege of delivering a talk on the joint work we conducted with Francisco Aliaga. The presentation was set for 12:15 pm, allowing us precisely 20 min to share our research. The title of our talk was "Fabrication of Stone Objects, by Geopolymeric Synthesis, in the Pre-Incan Huanka Civilization, Peru." (Davidovits and Aliaga 1981).

As the congress drew close on Tuesday, around 1:00 pm, my colleagues and I enjoyed a relaxed lunch. During this time, a man introduced himself as Walter Sullivan, a science journalist from the renowned New York Times. Intrigued by our presentation, Walter sought more information about our research. I later discovered that he was highly respected in the United States, and I attached great importance to our conversation on that memorable occasion.

Walter Sullivan's article commanded an entire page in the Sunday edition of the New York Times, dated May 24, 1981. The prominence of this placement, coupled

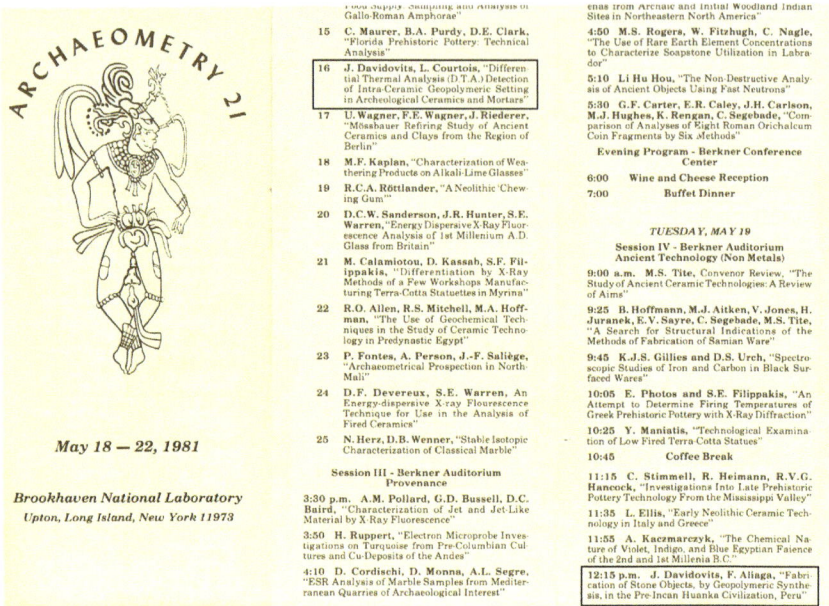

Fig. 1.1 Program of Archaeometry 21: Poster No. 16 and lecture on Tuesday, May 19, 1981 at 12:15

with an attention-grabbing advertisement, indicated that a significant number of readers would be captivated by its contents. The article bore the title "Testing of Relics Results in Surprises," offering a compelling glimpse into our research findings. It was not only the New York Times that took notice; other newspapers across the United States also picked up the story, albeit with a different title: "Technology unravels old riddles" (see Fig. 1.2):

> THE NEW YORK TIMES, SUNDAY, MAY 24, 1981, page 39. Testing of Relics Results in Surprises, By Walter Sullivan. After presenting the conference and some archaeological topics, W. Sullivan discusses our research project:
>
> (…) Probably the most sensational proposal at the meeting was that many of the most impressive ancient monuments of the world were not carved but cast from stone converted into a plastic form by plant extracts, such as oxalic acid, found abundantly in rhubarb leaves. Examples cited included stones forming the pyramids of Egypt. the ancient Beetle-browed statues of Easter Island and the great stone structures of the high Andes, such as the famous Gate of the Sun built by the ancient Huanka civilization at Tiahuanaco.
>
> The Work of Witch Doctors
>
> The proposal was made by Joseph Davidovits of the Geopolymer Institute in Saint-Quentin, France, who has been working with Francisco Aliaga, a Peruvian ethnographer. He pointed to a grove in the stonework of the Gate of the Sun that could have been produced by a fold in the rubberized cloth onto which the plastic rock had been poured. It has recently been discovered, he said, that some witch doctors, in the Huanka tradition, still make small stone objects in this manner. He cited evidence of oxalic acid derivatives in such monuments as those on Easter Island. Others have attributed these oxalates to fungi or micro-organisms.

> For a number of years Davidovits has argued that ancient ceramics were produced by a cementing process without any need for firing. His proposals were greeted with skepticism. "Intriguing, but definitely controversial," said Edward V. Sayre of Brookhaven, one of the conference organizers. Others argued that the use of heat to produce pottery from earliest times was well documented. (…)

The broad coverage of our work underscored its impact and generated widespread intrigue.

1.4   Back Home, the Research Continues

Upon returning home from the whirlwind of activity and the trip to the United States, our research endeavors continued. The primary focus was gathering as much scientific data as possible to extract organic acids from plants. Engaging in parallel work on industrial applications of geopolymers has proven to be a blessing, as it connected me with individuals possessing diverse scientific knowledge and unique perspectives. One such individual was André Bonnet, a promising young scientist from the Pharmacognosy Laboratory at the Faculty of Pharmacy in Grenoble (France).

I approached André with a proposal to conduct a study on the subject. While he discussed this possibility with his professor in Grenoble, we started our own investigation at our laboratory. Our initial study revolved around the dissolution of limestone, a material highly susceptible to acid attacks. I was well aware of limestone's vulnerability to acids, but what intrigued me was determining which specific organic acid, or combination of acids, would prove most effective. Surprisingly, no one had explored this line of inquiry before.

To initiate our exploration, we decided to employ the acids readily available in our laboratory: acetic acid (vinegar), citric acid (orange), and oxalic acid (rhubarb). Through a series of meticulous tests, we discovered that a combination of all three acids exhibited the most extraordinary prowess in disintegrating limestone.

In the meantime, the Faculty of Pharmacy agreed to undertake a study on the various acids that are extractable from plants. This collaboration promised to expand our understanding of the subject matter and shed light on the potential applications of these organic acids.

1.5   The Congress in Bradford, England, U.K. In 1982

Following the successful congress at the Brookhaven National Laboratory in New York, our next venture into the world of Archaeometry took us to the University of Bradford in England in March/April 1982. I had the privilege of presenting a paper alongside my colleagues titled "The Disaggregation of Stone Materials with Organic Acids from Plant Extracts, an Ancient and Universal Technique" (Davidovits et al. 1982). Our study led us to a fascinating conclusion: "… pre-Colombian farmers

Fig. 1.2 Walter Sullivan's article in the New York Times of May 24, 1981. In the center, the exact text was reprinted in a newspaper in Los Angeles, California

possessed the knowledge and ability to produce significant quantities of acids from regionally abundant plants, such as fruits, potatoes, maize, rhubarb, rumex, agave americana (yes, the cactus), ficus indica, oxalis pubescens, and more."

Our research revealed that these organic acids facilitated its decomposition when applied to limestone, resulting in a clay-like paste. However, we faced a significant challenge: how could we transform this paste into a durable rock that could withstand the elements and resist water? This remained a perplexing puzzle, and our knowledge fell short. To find answers, we knew we had to venture further, conducting additional research and tests. We did not know it would take us 34 years to uncover the missing piece of the puzzle.

During this time, I decided to shift my focus away from the research on pre-Columbian monuments of South America and dedicate myself entirely to studying industrial applications of geopolymers. However, my scientific curiosity remained insatiable, leading me to go deeper into the mysteries of ancient Egypt and its pyramids.

In 1997, with the advent of the Geopolymer Institute's internet platform, we seized the opportunity to share our findings with the world. We made the articles from the New York and Bradford congresses publicly accessible on our website under the title "Making cement with plant extracts." These articles, available in our digital library, have been freely downloaded over 10,300 times between 1997 and 2020. They have been referenced in various publications, popular science books, and online resources, reaching a broad audience and sparking further interest in this subject.

1.6 Waking Up in 2016: Ralph's Visit to Easter Island

It was June 2016. My son Ralph, who works with me at the Institute, told me of his plans to travel to Chile. He wanted to visit a childhood friend who settled in South America with his family. At last, after so many years, I thought this was an excellent opportunity for Ralph to take the time to visit Easter Island and thus resume our old "business."

With Ralph's planned trip in November, I had ample time to delve into my old archives and locate my books on the subject. However, my immediate focus shifted towards preparing for our annual Geopolymer Camp congress, which is held at the University Campus in Saint-Quentin, where our laboratory is affiliated.

On Sunday, November 13, at 2 pm on Easter Island and 9 pm in France, I eagerly awaited my videoconference connection with Ralph, who had arrived on the island a day earlier. Despite facing signal limitations preventing a video conference, we established an audio link. Ralph had rented a small scooter to explore the island swiftly. On his first day, he visited two significant archaeological sites: Vinapu, the closest to town, and the world-famous Ahu of Tongariki, with its impressive lineup of 15 statues.

The following day, Ralph returned to Vinapu. Our fascination with this site stems partly from the ongoing controversy that emerged after Thor Heyerdahl's Kon-Tiki expedition in 1947. Heyerdahl famously crossed the Pacific Ocean from Peru to the west on a raft. It is said that visitors to Vinapu are astonished by the resemblance of its walls to ancient structures found in Peru, particularly in the Cuzco area or in Tiahua-naco, Bolivia. After his visit, Ralph informed me that the similarity between the two constructions was strikingly evident (refer to Fig. 1.3). Heyerdahl's expedition leaned towards the hypothesis that the initial migrations to Easter Island originated from South America, coming from the east. Subsequently, the island would have been colonized by people from the west, specifically from Polynesia.

Upon revisiting the work of Francis Mazière, whom I mentioned earlier in this chapter, I share a quote from his book "Fantastique Ile de Pâques" regarding Vinapu, found on page 59 (English translation from French): "(…) it is true that the Ahu

Fig. 1.3 One of the Vinapu walls: **a** Ralph in front of the wall (Vinapu 1). **b** The white arrow in the upper middle part indicates the so-called "key" of stability, characteristic of the pre-Inca constructions of Cuzco in Peru

Vinapu represents a departure from Polynesian architecture [compared to the rest of Easter Island]… however, we cannot ignore the peculiar nature of this Vinapu architecture, which, under specific criteria, bears a resemblance to pre-Inca constructions (…) The evidence may be subtle, but it cannot be dismissed outright. Even if certain viewpoints challenge Heyerdahl's stance, we cannot disregard everything a priori (…)."

Indeed, when it comes to the walls, the construction techniques may exhibit similarities or even outright identity with those found in South America. However, what about the statues, the renowned Moai that stand gazing into the infinite? Could there also be a connection to the civilizations of the East in these enigmatic figures? It appears that the Heyerdahl expedition remains a sensitive topic for the island's inhabitants and is not entirely open for discussion. Ralph confirmed this in one of his emails from the island, mentioning, "I just visited the museum. I captured photographs of all the panels (around twenty in total). They solely focus on the Polynesians who arrived between [AD] 800 and 1200. No other hypothesis is entertained for a moment. They provide a comprehensive explanation of Polynesian culture, religion, navigation practices, rituals, and more. There is only a brief mention, when discussing the local flora and the plants introduced by the Polynesians, about the Chilean palm tree: some believe it to be the same species, while others consider it a derivative species… It is more of a museum centered around the Polynesian culture of the island rather than the Moai (the statues) that are an integral part of this culture." Ralph could not find a single word about the Vinapu walls.

In the exploration program I drafted for Ralph, a crucial element of his visit is a specific location called "Poike's Trench," which forms a depression. Theoretically, it may be possible to discover traces of intentional clay burning within this pit. Clay would constitute one of the primary ingredients for creating the geopolymeric binder used to construct the statues.

On November 15, Ralph shared an exciting update from the island, stating, "I'm in the pit, and I can see burnt clay on one of the faces. I'm collecting several samples." The other essential raw material could be the sweet sap of the abundantly growing Chilean palm trees. Recent studies have counted over 6 million of these trees on the island. The sap from these trees might have been used to produce vinegar through natural fermentation, utilizing the organic acid, acetic acid.

On this very day, November 15, 2016, I sent Ralph a message expressing my concern and the impasse I had encountered 34 years ago, leading me to halt all progress after the Bradford Congress in 1982. Though my message was brief, its significance was paramount:

> The acid tests are progressing well. After 20 days, with vinegar alone, it solidifies. After 3 days, the mixtures of vinegar + lactic acid + citric acid begin to solidify. It remains covered (without drying) [similar to a geopolymer binder].

During my lecture at the New York Congress 35 years ago, I raised the possibility that the Easter Island statues could have been artificially created. Access to samples from these statues would be necessary to prove this theory. Unfortunately, such samples do not exist. In fact, all my extensive research on the subject has been in vain, yielding no results. Easter Island may be an open-air museum, but like any museum, obtaining permission to extract even the smallest sample is impossible. The sole study conducted in 1972 by UNESCO remains my only available resource.

1.7 Early Travelers Describe Statues Made of Artificial Stone

Nevertheless, there are numerous intriguing clues to consider. It is widely known that Easter Island was discovered in 1722 by a Dutch expedition led by Jakob Roggeveen. In his logbook, he wrote:

"April 5, 1722: We spotted a turtle, followed by floating vegetation and birds. In the late afternoon, the Africanansche Galey [one of the expedition's ships] sighted land. We decided to anchor and wait until the next day. Since it was Easter day, Roggeveen decided to name the island Easter Island."

Now, let us delve into the more technical aspects concerning the statues. To ensure accuracy, I will provide the original text as well as the correct English translation. It has come to my attention that many translations or quotations cited by various authors do not align with the actual content of the original text. It seems these authors have misunderstood the technical terms used. Let me share Roggeveen's account, as presented by Cauwe and de Poorter (2020, p 226):

"10 April 1722: (…) These stone statues caused astonishment on our part, because we could not understand how it was possible that these people, without having at their disposal large and thick trunks to make some kind of machine, nor solid ropes, could erect such statues of at least 30 feet high and of fairly wide proportions. However, this perplexity ceased with the discovery, when detaching a piece of a statue, that it was made of clay or rich earth and that small flints had been pushed into it afterwards and very close together so as to prepare a figure of a man (a statue)". Text in Old Dutch: Deze steenen beelden hebben in 't eerst veroorsaekt, dat wy met verwondering aengedaen wierden: want wy konden niet begrypen hoe 't mogelyk was, dat die menschen, die ontbloot syn van swaer en dik hout om eenige machine te maeken, mitsgaders van kloek touwwerk, echter soodanige beelden, die wel 30 voeten hoog en naer proportie dik waren, hadden konnen oprigten: doch dese verwondering cesserde met te ondervinden door het aftrekken van een stuk steen, dat deze beelden van kley ofvette aarde waeren geformeerd, en dat men daerin kleene gladde keysteentjes hadde gestooken, die heel digt en net by den anderen geschikt synde, de vertooning van een mensch maekten.

Some 52 years later, in March 1774, James Cook made an entry in his diary (Cook 1777): (…) I had an opportunity of examining only two or three of these statues, which are near the landing-place; and they were of a grey stone, seemingly of the same sort as that with which the platforms were built. But some of the gentlemen who travelled over the island and examined many of them, were of opinion that the stone of which they were made, was different from any they saw on the island, and had much the appearance of being factitious [artificially created]. We could hardly conceive how these islanders, wholly unacquainted with any mechanical power, could raise such stupendous figures and afterward place the large cylindric stones before mentioned upon their heads. (…) But if the stones are factitious, the statues might have been put together on the place, in their present position, (…) But, let them have been made and set up by this or any other method, they must have been a work of immense time, and sufficiently shew the ingenuity and perseverance of these islanders in the age in which they were built. (…).

Both Roggeveen and James Cook, the first explorers to document the island, believed these statues were crafted from artificial stone or a material that imitated stone. In the case of James Cook, he was accompanied by two "naturalists," scientific experts knowledgeable in plants and rocks, thus lending credibility to their opinions. They acknowledged the possibility of the rock being artificial. This idea, which may seem heretical in our current time, appeared perfectly reasonable to them. It is important to remember that this was 1722 and 1774, respectively, a period that marked the end of the Baroque era in Europe. Palaces and churches were adorned with decorative elements in the "rococo" style, including stucco, simulated stone, and artificial columns. Encountering such materials on this remote island at the world's edge did not appear surprising.

However, these perspectives started to shift during the nineteenth century when other travelers discovered the unfinished statues scattered on the slopes of the Rano Raraku volcano. The transport methods of these massive statues continue to puzzle people. Fortunately, ethnologists like the Belgian Nicolas Cauwe are beginning to demystify the myth surrounding statue transportation. We will explore this topic further in Chaps. 11 and 12.

Now, armed with the knowledge we have gained from the science of geopolymers, it becomes increasingly clear to me that the statues were created using the technology of malleable stone. This conviction is reinforced by the South American tradition in Peru and the intriguing discoveries made by anthropologist Francisco Aliaga, which hint at a potential connection between various civilizations. To establish this interrelationship, our first step is to uncover and scientifically elucidate the historical context of the builders responsible for the awe-inspiring pre-Columbian megaliths found in the Altiplano region.

To commence this crucial endeavor, we must carefully select a site that offers convenient access to sample collection or has undergone some level of geological and mineralogical study. Fortunately, we have discovered an archaeological site that fits these criteria and, perhaps by chance, has remained relatively unknown to the general public. This advantageous circumstance will facilitate our initial investigations and provide a fertile ground for our research. In the forthcoming chapters, I will explore the remarkable findings we have uncovered at the site of Pumapunku, located in Tiahuanaco, Bolivia, where our journey into this captivating realm first began.

References

Cauwe N, de Poorter A (2020) La fabrique de l'histoire à Rapa Nui. Le transport des statues et l'ethnoarchéologie, des cas d'école. In: Dotte-Sarout E et al (eds) Pour une histoire de la préhistoire océanienne, Cahiers du Credo, chap. 10. Aix-Marseille Université, p 226

Cook J (1777) Voyage towards the south pole and round the globe. Book I, Chapter VIII, p 295. Printed for W. Strahan and T. Cadell, London

Davidovits J, Aliaga F (1981) Fabrication of stone objects by geopolymeric synthesis in the Pre-Incan Huanka civilization, Peru. Abstracts of Archaeometry 21. Brookhaven National Laboratory, New York, USA, p 21

Davidovits J, Bonnet A, Mariotte AM (1982) The disaggregation of stone materials with organic acids from plant extracts, an ancient and universal technique. In: Proceedings of the 22nd symposium on archaeometry. University of Bradford, UK, Mar 30–Apr 3 1982, pp 205–212

Chapter 2
Why Did We Choose Pumapunku/ Tiwanaku and Not Cuzco and Sacsayhuaman?

Why did we choose Pumapunku/Tiwanaku and not Cuzco and Sacsayhuaman?—
We set up our exploration team—Why not choose the Sacsayhuaman Fortress in
Peru?—The scientific study of the Pumapunku red sandstone—Another alternative
in Peru?

The decision to focus our research on Pumapunku/Tiwanaku rather than Cuzco
and Sacsayhuaman was influenced by various factors inherent to the scientific field.
As a scientist, I had to consider these factors and make informed choices accordingly.

In September 2017, my second son, Frédéric, was in Paris at the Bibliothèque du
Musée des Arts Premiers, diligently scanning each of the 150 pages of a significant
scientific report for me. This report provided a comprehensive petrographic analysis, encompassing mineralogical and geological aspects, of the rocks utilized in the
construction of the monuments at the Pumapunku site in Tiahuanaco, Bolivia. The
mission assigned to Frédéric was to bring back a complete copy of this work, as it was
the only existing specimen in France. The report, titled (Translation from Spanish)
"Provenance of the Sandstones Used in the Pre-Columbian Temple of Pumapunku
(Tiwanaku)," was authored by Carlos Ponce Sangines, who served as the Director of
the Archaeological Research Centre at Tiwanaku (as depicted in Fig. 2.1). It is worth
noting that while the site is commonly spelled Tiahuanaco, archaeologists prefer to
use Tiwanaku when referring to the historic site.

This invaluable report played a significant role in shaping our research direction.
The depth of analysis and insights it offered regarding the specific rocks present
at Pumapunku/Tiwanaku made it an ideal starting point for our investigations. The
information contained within this work served as a solid foundation upon which we
could build our scientific endeavors, providing a comprehensive understanding of
the materials employed in the construction of the remarkable structures at the site.

While Cuzco and Sacsayhuaman undoubtedly possess their own intriguing
mysteries and historical significance, our decision to prioritize Pumapunku/Tiwanaku

© The Author(s), under exclusive license to Springer Nature Switzerland AG 2024   15
J. Davidovits, *Ancient Geopolymers in South America and Easter Island*,
SpringerBriefs in Earth Sciences, https://doi.org/10.1007/978-3-031-75336-7_2

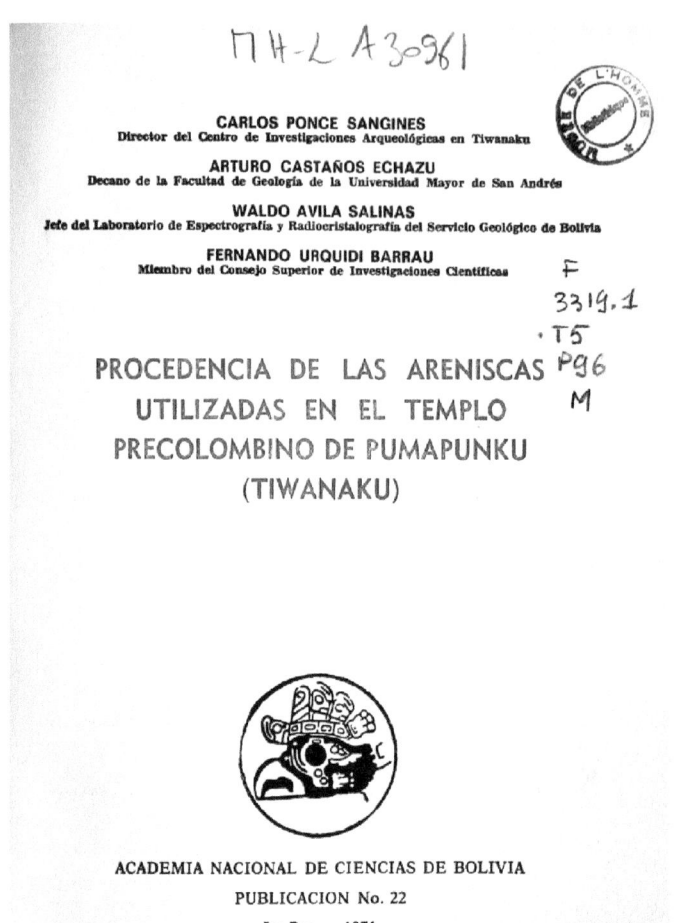

CARLOS PONCE SANGINES
Director del Centro de Investigaciones Arqueológicas en Tiwanaku

ARTURO CASTAÑOS ECHAZU
Decano de la Facultad de Geología de la Universidad Mayor de San Andrés

WALDO AVILA SALINAS
Jefe del Laboratorio de Espectrografía y Radiocristalografía del Servicio Geológico de Bolivia

FERNANDO URQUIDI BARRAU
Miembro del Consejo Superior de Investigaciones Científicas

PROCEDENCIA DE LAS ARENISCAS UTILIZADAS EN EL TEMPLO PRECOLOMBINO DE PUMAPUNKU (TIWANAKU)

ACADEMIA NACIONAL DE CIENCIAS DE BOLIVIA
PUBLICACION No. 22
La Paz — 1971

Fig. 2.1 Cover of the scientific report on the origin of the sandstone that forms the monuments of Pumapunku by the team led by Ponce Sangines (1971)

was driven by the availability of this exceptional scientific report and its potential to guide our research in a focused and informed manner.

As I mentioned in the previous chapter, our research journey necessitated the identification of South American civilizations that had implemented the technology of malleable stone, as indicated by the traditions of Peru and the ancestral knowledge passed down to anthropologist Francisco Aliaga. Initially, my focus was drawn to the remarkable monuments like the zigzag wall of Sacsayhuaman and others that were more widely recognized. However, I understood the importance of widening our search and delving into the work of Carlos Ponce Sangines, which I eagerly anticipated reviewing.

TRUEMMERSTAETTE VON PUMAPUNGU. ANSICHT VON NORDWEST

Fig. 2.2 The ruins of Pumapunku, photographed by a German expedition in 1876 (Stübel u. Uhle: Tiahuanaco; photograph by G. W. Grumbkow)

Pumapunku, an archaeological site known to only a select few, had been viewed in the past as nothing more than a field of ruins, as depicted in Fig. 2.2. However, I firmly believed that the work carried out by Carlos Ponce Sangines and his team held the key to providing me with a convincing answer and compelling reasons to guide my decision. It was a book published nearly 50 years ago, in 1971, and it became evident that certain things require time to unravel and reveal their hidden truths.

With great anticipation, I went into the pages of this book, fully aware that it held the potential to uncover invaluable insights and propel our research forward. The depth of knowledge and meticulous research within its chapters would undoubtedly provide me with the clarity and conviction I sought. Indeed, time had proven that patience and thorough investigation were essential to unraveling the mysteries that lay before us.

When I found the photograph taken by a German team in 1876, which I have included as Fig. 2.2, I had high hopes of discovering what I had envisioned countless times before—the remnants of an ancient temple wall or at least fragments of its structure. However, to my surprise, the image revealed something entirely different. It displayed the remains of megalithic terraces arranged in a captivating monolithic floor, with each individual component weighing over 100–130 tons. At first glance, these slabs appeared to be constructed from a type of "concrete," resembling the sedimentary rock red sandstone (as seen in Fig. 2.3).

As I studied this photograph and considered the unique characteristics of this stone, it became evident that our geopolymer technology could easily replicate this material artificially. The prospect of reconstructing these remarkable structures using our advanced techniques fueled my conviction that the study of Pumapunku's ruins was a pursuit worth undertaking.

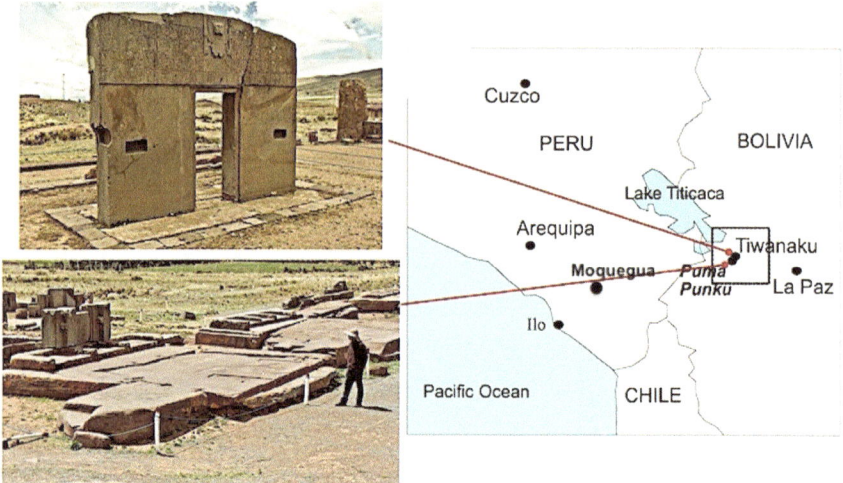

Fig. 2.3 Above, left: Tiwanaku and the Sun Gate; below, left: view of Pumapunku and its red sandstone megalithic terraces. Right, location map

However, before embarking on this research project, I had to follow a self-imposed requirement. In all of my archaeological studies, it has been my practice to rely on the works and insights of my scientific colleagues, using their research as a foundation upon which I can construct and design my own investigative plan. Even if I later discover that their findings are only partially valid, I find great value in their contributions. In this regard, the publication of Carlos Ponce Sangines and his team of scientists holds immeasurable worth to me. Their results and discoveries provide a solid starting point, allowing me to build upon their findings and shape my research trajectory.

2.1 We Set Up Our Exploration Team

Once we started planning the fieldwork and with the certainty of Ralph's trip to South America, Frédéric and I agreed that we could not let Ralph go alone to Pumapunku/Tiwanaku. Finding a local partner became a priority. I turned to my trusty computer for a solution. Did we have any previous connections with a local university? Yes, indeed! During Geopolymer Camp 2016, our annual international conference was held at the Saint-Quentin University Campus every July, and we had the pleasure of hosting a participant from Peru. He came from the Universidad Catolica San Pablo in Arequipa, the second-largest city in Peru after Lima. The Peruvian scientist Rizal Nuñez Dara was working on the applications of geopolymers in the transformation of mine tailings, which are the mineral residues from the extraction of copper ore. Copper mining is an essential economic activity in the southwest of Peru, in the

Arequipa/Moquegua Region. His studies aimed to transform these rocks and minerals into construction materials. I sent him the following e-mail:

"Bonjour, I hope you enjoyed attending and applied your knowledge you acquired in Geopolymer Science and Technology, at the 2016 Geopolymer Camp and Tutorial. Are you still in Arequipa, San Pablo Catholic University?

I am contacting you because we have started an archaeological project dedicated to the civilization of Tiwanaku. We have gathered very interesting information, reading numerous scientific and archaeological papers. We came to the conclusion that we need to have a better knowledge on the geological resources of two precise sites:- one involving the Moquegua Valley not far south of Arequipa (your region) down to the Pacific Ocean (Ilo);- one involving the Tiwanaku region itself, Bolivia. We are looking for clays and weathered stone outcrops for sandstone and andesite.

Therefore, my son Ralph has decided to make a 7–10 day visit to these sites, and we wonder whether you could be available to accompany him during this trip. Nearest schedule would be the end of October-beginning of November this year. Of course, all expenses would be at our charge. Ralph could give a seminar/talk on geopolymers at your university. If you are available, it would be perfect. Looking forward to your comments, JD."

The response did not keep us waiting for long. Two days later, an e-mail from Saul Perez, Head of Research and Science Projects at the Catholic University of San Pablo (UCSP), landed in my inbox. Saul, who was also part of Professor Huaman's research team, expressed great interest in the possibility of a presentation at the university and showed eagerness to collaborate with our research in any way he could. In addition, he informed me that Luis Huaman, the professor's brother and a geology student, would be more than willing to accompany Ralph on the field visit.

Excitement filled the air as things started falling into place. The UCSP agreed to support us and graciously assigned Luis Huaman, a young geology student with a background in geopolymer science, to our team. He was a significant asset since geologists typically focus on the natural components and elements found in the rocks they study. It was crucial for Luis to witness firsthand that synthesizing certain minerals is a relatively straightforward process that does not require extreme pressure, high temperatures, or millions of years. While it is not common to find geologists open to the new science of geopolymers, more and more individuals like Luis are embracing this field every day, and we were fortunate to have him on board. It was clear that we were heading in the right direction.

As we continued to organize the logistics of the trip, we came to the conclusion that Arequipa would serve as an excellent base for our expedition. Situated at an altitude of 2000 m above sea level, it would allow Ralph to acclimate to the altitude before venturing into the challenging conditions of the Altiplano, where he would be at an elevation of over 3800 m.

2.2   Why Not Choose the Sacsayhuaman Fortress in Peru?

In the earlier part of this chapter, I emphasized the significance of the research conducted by Carlos Ponce's team, who conducted a meticulous petrographic study at the site of Pumapunku. Despite my exhaustive search, I could not find any similar research material for archaeological sites in Peru. As I looked into the bibliography of the renowned Sacsayhuaman site (Fig. 2.4), it became apparent that Peruvian archaeologists attributed the fortress to the Incas, dating its construction to AD 1300–1400, a mere 600 years ago, placing it in the same timeframe as Christopher Columbus' discovery of America.

However, it is important to note that my previous statement is not entirely accurate. The dating of Sacsayhuaman's construction was primarily conducted by anthropologists and historians from Europe and North America who, until recently, visited various archaeological sites with their students. They would spend several months conducting field research, financially supported by their respective universities. Additionally, researchers from France, Germany, Italy, the United Kingdom, and other countries operated independently, focusing their studies on specific archaeological sites, with their research funded by their home countries. In this context, the host country primarily provided local labor, boosting the local economy.

Nevertheless, let us revisit the example of Sacsayhuaman: the fortress that reveals distinct stages in its structure to every visitor. The first wall consists of colossal megaliths, some weighing up to 350 tons, positioned at ground level. Moving up, we encounter the second and third terraces, constructed with smaller blocks. According to certain accounts, the larger blocks were created by a civilization predating the Incas. It was the Incas who, centuries later, added the upper terraces, intriguingly composed of smaller blocks. Given this information, it would be fascinating to investigate whether the mineralogical composition of both ancient and more recent block

Fig. 2.4 General view of the Sacsayhuaman site near Cuzco

types is the same. Ideally, such a study would have been conducted by Peruvian scientists initiated and supported by the Ministry of Culture. It would involve a comprehensive examination comparing both block types' mineralogical and geochemical structures, supplemented by geological research to identify the quarries where these materials were sourced and extracted. Regrettably, such a study does not exist.

However, there are intriguing clues scattered throughout scientific publications that I studied, thanks to the compilation efforts of the American amateur archaeologist Kon Khanyants, to whom I am grateful for his assistance. Through his collaboration, I have found articles like those written by Carlos Kalafatovich (Kalafatovich 1970), a Geology professor at the University of Cuzco, published in 1957 and 1970. In his work, he highlighted, for instance, in the Revista del Patronato Departamental de Arqueologia del Cuzco in 1970 (translated from Spanish):

"... As for the quarries used for the construction of the walls of the Fortress, it is clear that they have used the limestone that crops out in more or less extensive masses, a few hundred meters to the north and east of this archaeological group, but many blocks would also have been extracted from the site of the Fortress, since, as can be seen in the geological map, there are limestone outcrops above the walls..."

Regrettably, the claims made in this article lack the support of petrographic, mineralogical, or geochemical studies. Likewise, other brief articles also assert that the limestone quarries are located at distances ranging from 3 to 15 km from the monuments, contradicting the conclusions of the Peruvian geologist. It becomes apparent that the composition of the megaliths is limestone, not andesite or other volcanic rocks, as commonly described in popular books. As is often the case, we are left with unreliable, partial, and incomplete studies. This lack of solid scientific foundation explains the emergence of alternative theories, as there is a dearth of conclusive evidence.

2.3 The Scientific Study of the Pumapunku Red Sandstone

Ironically, in the same year, the Bolivian government made a contrasting decision by actively engaging its scientists in a thorough and meticulously documented examination of the Pumapunku site at Tiwanaku. Their efforts culminated in a research report Frédéric was tasked with bringing back to Saint-Quentin. Within this report, Carlos Ponce Sangines eloquently describes the objective of his project as follows:

"Purpose of the present monograph: It basically involves two purposes:

(1) To encourage the conviction that the enhancement of the ruins of the pre-Columbian temple of Pumapunku is an unavoidable imperative and should therefore be included in the programme of activities of the Centre for Archaeological Research in Tiwanaku within a reasonable period of time. The term "enhancement" means to provide it with objective and environmental conditions, which, without disfiguring its nature, make it possible to highlight its specific characteristics and allow it to be used in the best possible way. (…) In this sense, the present work, on the one hand, tends to revalue and draw attention to Pumapunku, which when restored will

occupy a pre-eminent place not only in Bolivia's monumental heritage but also in South America's. No effort will be in vain (…).

(2) The second purpose of this book is to determine the exact place from which the red sandstone used in the lithic of the stone platform of the pre-Columbian temple of Pumapunku was extracted. (…) The step to be taken is straightforward. It lies in the comparison of thin sections, observed by petrographic microscopy, with the intention of establishing whether the rock samples of the ancient specimens are similar or dissimilar to those from known quarries (…).

(3) Samples from Pumapunku, lithic platforms: 20;- Samples from the southern highlands: 27; Total samples analyzed: 47…."

As part of the accomplished team that contributed to the aforementioned objective outlined by Carlos Ponce, we had the presence of Arturo Castanos Echazu, responsible for conducting petrographic analysis; Waldo Avila Salinas, who oversaw the X-ray refraction examination; and Fernando Urquidi Barreau, who led the geochemical study. Their conclusions were derived from a meticulous investigation involving a substantial number of samples. They collected 20 samples from the four red sandstone platforms and an additional 27 samples from six carefully chosen geological sites located south of Tiahuanaco, specifically in the Quimsachata Mountain, which is situated at distances of 8 and 11 km away from Pumapunku. The intriguing findings from this research will be thoroughly explored in Chaps. 6 and 8.

2.4   Another Alternative in Peru?

During my research, I came across the work of American archaeologist Paul Goldstein, who published an article in 1993 titled *"Tiwanaku Temples and State Expansion: A Tiwanaku Sunken-Court Temple in Moquegua."* (Goldstein 1993). Goldstein's research revealed that at the OMO site in Moquegua, the Tiwanaku people had constructed a temple bearing all the distinctive characteristics of those found in the Altiplano. The town of Moquegua is situated in a valley that descends westward towards the Pacific Ocean (refer to the map in Fig. 2.3). The article states:

"(...) Great care is also evident in the construction of the Upper Court superstructure, where a single course of finely worked ashlar blocks was used as the foundation for most interior walls. Because of the intentional destruction of walls, the foundations of the Upper Court superstructure were only found intact in areas buried under particularly deep layers of adobe wall-fall. (...) The ashlar blocks were all rectangular, although sizes varied considerably. Three or more faces of each ashlar were extraordinarily well smoothed by abrasion, while non-visible sides were left unpolished. The vertical edges of each stone were closely fitted to its neighbors without any mortar or space between blocks. Ashlars in the same course varied in height and length. This technique is very similar to the distinctive masonry of the Pumapunku Temple (...)"

Upon reading Goldstein's work, I initially believed that there was no need to venture to Bolivia, as Moquegua, a mere hour and a half's drive from Arequipa, presented us with building blocks resembling those of Pumapunku, crafted from red sandstone. However, my initial excitement dwindled as I continued my research, particularly when I came across an article in La Republica newspaper titled *"Moquegua: Imposing Temple of Omo."* Penned by Karen Rodríguez Carpio and dated November 25, 2014, the article conveyed the following information (translated from Spanish) (Karen Rodríguez Carpio 2014):

> Arriving at the Omo sector, located in the lower part of the Moquegua Valley, was an odyssey. A wild hare and an eagle gave me a "welcome"; I did not know whether to return to the vehicle that transported me to the place or continue appreciating our rich past, so I weighed my heart and continued there. Moquegua is home to dozens of archaeological sites that, due to a lack of a budget, either from the central government or private enterprises, we still cannot enjoy. This is the case of the "Temple of Omo"; excavations began in 1990 and continued until 2013. Funding was provided by the University of San Diego (USA), appointing archaeologist Paul Goldstein, who specializes in the Tiahuanaco culture, as project director. In 2004, he presented his book "Andean Diaspora. The Tiahuanaco, colonies and the Yens of South America", in which he illustrates the study of this important temple.
>
> The project's co-director in Peru, Patricia Palacios, pointed out that the investigation of the religious structure was completed in its entirety, lacking only the results of sampling cemeteries adjacent to the temple. These cemeteries were located by an archaeological survey between 1996 and 1998 in the lower part of the valley of Moquegua. In the excavations carried out by a group of North American archaeologists, it was found that the temple built between AD 500 and 600, in an area of 400 m^2, is divided into three levels: the lower patio, where the population could enter without any restrictions; the middle patio, where there were certain restrictions as there were two doors, and the upper patio which had one door, allowing only the passage of the religious dome. (...) The Omo Temple was destroyed by the same population between 800 and 900 AD due to political problems. Palacios said that the temple is currently protected by layers of earth because the climatic conditions in the province are not conducive to its maintenance. In order to be able to appreciate it as part of a tourist route, the collapsed walls need to be rebuilt. (...) In the meantime, the people of Moquegua will have to wait a long time to appreciate the imposing "Temple of Omo."

Upon receiving this new information, I promptly contacted our partners in Arequipa to gather more details regarding the recent findings mentioned in the press. We consulted Luis Huaman, who provided us with a comprehensive report on the matter on October 26, 2017. According to Luis, after conversing with Patricia Palacios, the curator of the Contisuyo Museum in Moquegua, it was confirmed that the M10 temple (as it is internally referred to) had been intentionally covered with sand for conservation purposes. Luis further explained that, upon its discovery, there were initial plans to transform the temple into a tourist attraction. However, due to the lack of necessary funding, allocating the resources required to prevent deterioration proved impossible. As a result, the M10 temple remains buried in its original state, though the Contisuyo Museum has preserved some sample elements of the temple.

Luis clarified that obtaining physical samples from the museum for academic purposes was impossible. All the materials are considered an archaeological heritage of Peru, and the process of obtaining a formal permit to conduct research at an archaeological site could become a lengthy and bureaucratic ordeal, potentially taking more than six months. Luis recognizes this frustrating succession of red tape as a tedious process.

With limited options, Ralph's only course of action would be to personally visit the museum, observe the stones, and capture photographs. However, Luis brought a new piece of information that dashed our plans. The museum's strict regulations prohibit any photography of the displayed artifacts. At this juncture, I believe that we have gathered enough evidence to dismiss Moquegua and channel our energy and research toward Pumapunku, located in the Bolivian Altiplano. After this arduous journey, filled with unexpected twists and necessary detours, we are ready to embark on our final destination in the south: Pumapunku, Bolivia.

References

Goldstein P (1993) Tiwanaku temples and state expansion: a Tiwanaku Sunken-Court temple in Moquegua. Lat Am Antiq 4:22–47

Kalafatovich C (1970) Geologia del Groupo Arqueologico de la Fortaleza de Saccsayhuaman y sus Vecindades. Revista del Patronato Departemental de Arqueologia del Cuzco 1:61–67

Karen Rodríguez Carpio La Republica (2014) https://larepublica.pe/archivo/836588-moquegua-imponente-templo-de-omo/

Ponce Sangines C, Castaños EA, Avila SW, Urquidi BF (1971) Procedencia de las Areniscas utilizadas en el Templo Precolombino de Pumapunku (Tiwanaku). Academia Nacional de Ciencias de Bolivia, Publication No. 22, La Paz

Chapter 3
The Preliminaries of the Expedition and the Question About the Extraterrestrials

Definition of the exploration program—Did the aliens build Pumapunku?—The official Bolivian view.

On November 5, 2017, Ralph arrived at Arequipa airport, where he was greeted by Luis Huaman, the young geologist who would be joining us on this expedition, along with Luis' brother Fredy Huaman, who is also a geologist and a professor at the Universidad Católica San Pablo (UCSP). After exchanging pleasantries and introductions, they proceeded to discuss and finalize the details of Ralph's stay in the city, as well as the upcoming expedition to Tiahuanaco.

The following day, the program was set to commence with a packed schedule. Ralph would kick off the day by delivering a presentation on geopolymer technology at UCSP in the morning. In the afternoon, he would give a lecture in English titled "Geopolymer Technologies: from Theory to Global Industrialization" in Spanish: Technologias Geopolymericas: de la Teoria à la Industrializacion Global (see Fig. 3.1).

3.1 Definition of the Exploration Program

Despite the well-planned schedule, Ralph found himself struggling with jet lag due to the 6-h time difference between France and Peru, as well as an overnight stopover in Lima. This sleep issue resulted in a state of fatigue that he had underestimated. In his correspondence with me, Ralph confessed that he had miscalculated the travel times, leading to the need for some adjustments to the schedule.

Originally, Ralph had intended to rent a car on Tuesday, November 7, to accompany Luis on a visit to Ilo, with a detour to Moquegua (as described in Chap. 2, Fig. 2. 3). However, taking into account local conditions, the journey to Moquegua alone would take a minimum of four hours, with an additional 2 h to visit the Ilo site, and

J. Davidovits, *Ancient Geopolymers in South America and Easter Island*, SpringerBriefs in Earth Sciences, https://doi.org/10.1007/978-3-031-75336-7_3

Fig. 3.1 Announcement of Ralph Davidovits' conference on November 6, 2017, at Universidad Catolica San Pablo, Arequipa, Peru

another four hours for the return trip, not accounting for any breaks. Considering this, during our previous video call, we mutually agreed that the journey to Moquegua could be postponed since it was not imperative for the purposes outlined in Chap. 2.

Furthermore, the appointment to visit the Ilo Reserve had been scheduled for 8 am, which would require leaving Arequipa at 4 am. To accommodate Ralph's fatigue and the revised schedule, we rescheduled the visit to Ilo after the Tiahuanaco trip, specifically on November 15. In the meantime, Luis would make the round trip from Arequipa to Ilo and back in a day, while Ralph would engage in technical discussions with the group working on geopolymer applications.

From my laboratory, I closely followed the tightly packed schedule and eagerly awaited the results of the visit to the Punta Colles National Reserve in Ilo. This excursion was of significant importance to our expedition, as we were searching for one of the essential chemical ingredients for the formulation of the "malleable stone": guano.

On Tuesday, November 7, I reached out to Saul, the professor in charge of research and development at UCSP, to inform him about the adjustments we made to our program due to unforeseen circumstances. I shared with him the updated itinerary and logistics that had been put in place:

Dear Saul:

Following my conversation with Ralph, I believe that he needs additional time to adequately deal with the difference in climate (altitude) and jet lag. Therefore, I do not think it is a good idea to travel overnight by bus from Arequipa to La Paz. I want him and Luis to be in good shape for the archaeological and geological expedition to the Altiplano. Therefore, I suggest that they take the option of travelling with a car and driver to the Bolivian border, and continue from there with a Bolivian driver or in a taxi. I will, of course, assume the costs involved.

We managed to rearrange everything so that Ralph and Luis could depart for Tiahuanaco on the morning of November 8. They would be accompanied by a driver who would take them from Arequipa to the Bolivian border at Desaguadero. From there, they would continue their journey to the hotel reserved in La Paz using a taxi. The expedition to Tiahuanaco and the geological sites would then take place from November 9 to November 12, allowing for a total of four full days to conduct this preliminary study.

Once they returned to Arequipa, Luis would make his way to Ilo on November 15. As for Ralph, he would be heading to Lima on November 16 to catch his flight to Chile on November 17. These were the adjusted plans to ensure the smooth continuation of our expedition.

3.2 Did the Aliens Build Pumapunku?

To fully grasp the intricacies and significance of our research, let me share a conversation that took place at the UCSP restaurant involving Ralph and a group of professors from different departments of the university. Prior to his lecture later that day, Ralph took the opportunity during lunch to explain the purpose of our proposed archaeological research. Our focus was the Pumapunku site, particularly the massive monolithic terraces made of red sandstone. Ralph discussed the possibility that these enormous blocks, weighing between 100 and 130 tons, could be composed of an artificial stone known as geopolymer concrete. Our expedition aimed to shed light on this hypothesis.

As Ralph elaborated on these details to the professors, they responded in a rather unexpected manner, referring to the popular belief held by many Peruvians that the monument was constructed by "extraterrestrials." Naturally, it became evident that our archaeological research had the potential to challenge established beliefs ingrained in the collective imagination. This scenario was not entirely unfamiliar territory for me, as I had previously encountered similar situations when studying the Great Pyramids of Egypt.

Months ago, while conducting online bibliographical research, I came across abundant information about the extraterrestrial theory surrounding Pumapunku. In fact, proponents of alien intervention in the construction of ancient archaeological sites consider Pumapunku to be the most compelling evidence. As an example, let me quote a passage from the French website Dramatic (translated from French):

> (…) If there is one place on Earth that shows traces of ancient extraterrestrials, it is Pumapunku. Are the ancient ruins of Pumapunku the result of the incredible ingenuity of primitive man? Or are they the product of a much higher power? Pumapunku is the sole site on planet Earth that some believe was built directly by aliens, and millions of people around the world think that aliens have visited us. What if this were true? Did ancient extraterrestrials really help shape our history? And if so, could there be evidence of a lost alien city in Bolivia? (…).

In the search for answers about Pumapunku, it seems that the tourism industry in Bolivia has embraced the idea that the site may have extraterrestrial origins (see Fig. 3.2 for reference). While official archaeology has yet to provide a conclusive answer, a simple search using Google reveals a plethora of information on the subject.

Among the numerous results, one account caught my attention—a blog post by a young French backpacker named Amandine, shared on her website Un sac sur le dos in 2014. Amandine recounts her adventures and experiences as she explores different destinations around the world. In her blog, she writes about Pumapunku, referring to it as the "site of the aliens." Amandine shares her curiosity about the existence of two sites to visit and obediently follows her guide to the second site. She describes her awe at the famous Pumapunku site and how it aligns with the local belief that aliens created it. Amandine's description implies that the site's unique characteristics lend credibility to this belief.

On the internet, there is a website called ancientcode.com that presents an alternative perspective on Pumapunku. This website does not explicitly mention extraterrestrials, but it does discuss a civilization with advanced technology. The article highlights 30 images of Pumapunku that supposedly demonstrate the use of advanced ancient technology. Located in the Andean mountains, just 45 miles west of La Paz, Pumapunku is described as a mysterious place with abundant megalithic stones, some of the largest ever found on Earth.

The article challenges conventional views on ancient cultures, emphasizing the precise craftsmanship, intricate cuts, and polished surfaces of the stones at Pumapunku. These characteristics have puzzled scholars for centuries. Pumapunku is part of a larger complex that belonged to the ancient Tiahuanaco culture, which predates the Inca civilization by several hundred years, if not millennia. The andesite stones used in construction fit together perfectly without the need for mortar. In fact, not even a sheet of paper can be inserted between these ancient stones. The article poses thought-provoking questions: How was such precision achieved thousands of years ago? Could this be evidence of otherworldly technology, as some suggest? Alternatively, could the ancient Tiahuanaco culture have possessed advanced technology that has been lost to history?

Fig. 3.2 Pumapunku; **a** the megalithic red sandstone slabs before the restoration in 1970; **b** current presentation of some andesite volcanic rock elements (2017)

Interestingly, the article mentions that the Inca tribes deny any involvement in the construction of Pumapunku, supporting the belief in ancient legends and disassociating their ancestors from the site. The stones at Pumapunku fit together like a giant puzzle, forming load-bearing joints that have withstood the test of time. While words can only convey so much, the article presents 30 captivating images of this enigmatic ancient complex. These images continue to defy explanation and challenge rational thinking.

3.3   The Official Bolivian View

Suppose we consider the existence of a significant group of Peruvian and Bolivian academics who support the extraterrestrial construction theory. It raises the question of how official archaeologists and scientists react to this perspective and what proposals they have put forth. In my research, I discovered that a conference called the VI Latin American Symposium of Physics and Chemistry in Archaeology, Art, and Conservation of Cultural Heritage (LASMAC) took place in La Paz, near Tiahuanaco, from 10 to July 14, 2017. This event was organized by the Center for Archaeological and Anthropological Studies and Administration of Tiwanaku (CIAAAT) with the aim of bringing together scientists and archaeologists from various Latin American countries, including Brazil, Chile, Argentina, Mexico, Colombia, Ecuador, Peru, and Bolivia.

During the symposium, many lectures focused on studies and research conducted by physicists and chemists in areas such as ceramics, paints, metals, fabrics, and more. However, I also came across a paper presented by two archaeologists and one anthropologist from the Universidad Mayor de San Andrés in La Paz, Bolivia, which caught my attention (refer to Fig. 3.3). The title of this work is "A contribution to Experimental Archaeology through the manufacturing processes in the making of the replica of the 'Fraile' Stela from Tiahuanaco" presented by Fernandez et al. (2017).

This paper suggests that the researchers aimed to contribute to experimental archaeology by replicating the "Fraile" Stela from Tiahuanaco and examining the manufacturing processes involved. While the details of their findings are not provided, it is intriguing to see archaeologists engaging in experimental approaches to gain insights into ancient techniques.

It is interesting to study further into the content of this paper and explore the perspectives and conclusions put forth by these Bolivian researchers. These archaeologists highlight that from the time of Cieza de León's initial writings (which I

Un aporte a la Arqueología Experimental mediante los procesos de manufactura en la elaboración de la réplica de la Estela el "Fraile" de Tiahuanaco

J.M. ....[1], S. D........[2], G.K. Y.......[3]
Email: m..........@hotmail.com

1) Egresado de la carrera de Arqueología, Universidad mayor de San Andrés
2) Egresado de la carrera de Arqueología, Universidad Mayor de San Andrés
3) Estudiante de la carrera de Antropología, Universidad Mayor de San Andrés

Fig. 3.3 Head of the lecture presented by Bolivian archaeologists at LASMAC 2017 by Fernandez et al. (2017)

will study further in Chap. 4) until the present day, there has been a consensus that we still do not understand how these massive structural elements were created. In the summary of their presentation, they emphasize: "(Excerpt of the abstract) The Tiahuanaco Culture has been the subject of numerous studies by both local and foreign researchers. The majority of scientific research took place in the latter half of the twentieth century, with most focusing on the monumental site in hopes of obtaining groundbreaking insights. However, fewer studies have been interested in studying aspects, such as the production tools for lithic and metallurgical processes, methods of sourcing raw materials, and other unresolved topics. (…) Cieza de León's observations prompt us to reflect on archaeology in Bolivia, as this new century presents an opportunity to offer a scientific contribution that aims to demonstrate the feasibility of creating large elements using small and medium-sized tools like hammers and polishers. These tools could have been employed in the construction of towering stelae in Tiahuanaco. (…) The objective of this project is to enhance our understanding of carving and polishing techniques by making comparisons between lithic objects used for percussion and polishing during pre-Hispanic times and the stone instruments used in similar work today. Many studies have attempted to unravel the mysterious "commercial" ideas that surround Tiwanaku, often overshadowing the scientific contributions related to the objects representing this Andean culture. Through the use of experimental archaeology, this project will provide yet another argument to demonstrate that the ancient Tiwanaku culture possessed the knowledge and skills to create the material remains that form part of the archaeological complex of Tiahuanaco and the wider socio-cultural context of Bolivia (end of excerpt)."

To go deeper into the analysis, let me provide some information: The "Fraile" statue is a 2.4-m-tall sculpture (refer to Fig. 3.4a), originally carved from a soft red and yellow sandstone with relatively flat and minimally emphasized high reliefs, making it less challenging to reproduce. It is precisely this monument that the archaeologists propose to replicate using simple tools such as stone hammers or strikers and abrasive sand.

If successful, according to their assertions, this endeavor would demonstrate that the ancient inhabitants of Tiwanaku possessed all the "primitive" means necessary to create the magnificent, enigmatic, and colossal megalithic blocks made of red sandstone. Consequently, it would also suggest their capability to fabricate other structural elements using the harder volcanic rock andesite.

Hence, the archaeologists made a deliberate choice not to work with a more challenging material like andesite or attempt to replicate a complex object such as the "H" shaped sculpture (refer to Fig. 3.4b). Instead, they opted for the easier-to-work red sandstone, a material commonly used in modern times for carving statues. I do not see this choice as a significant challenge since it involves no technological or technical difficulties. Reproducing the "Fraile" statue, in my opinion, can be accomplished without major complications or demands. However, it is worth noting that this group of archaeologists is participating in a conference where highly sophisticated technologies are being presented, including carbon-14 dating analysis for organic elements, thermoluminescence for ceramics, and research on various aspects such as ceramics, pigments, fabrics, paints, and metallurgy.

Fig. 3.4 **a** On the left, the "Fraile" statue in bicolor sandstone, which is easy to sculpt; **b** on the right, the H-shaped sculptures in alleged volcanic andesite rock, which are hardly impossible to sculpt (2017)

All these technologies, mastered to perfection by the Homo sapiens of the Altiplano, clearly demonstrate the deep knowledge these civilizations possessed regarding their mineral and plant resources within their immediate ecological environment. It is indeed curious that these anthropologists, during their research, do not question why and how the Homo sapiens of the Altiplano invented and developed these technological innovations, despite requiring extensive knowledge in "alchemical" and mineralogical realms. For instance, how did they choose one particular blue-colored mineral over another? What flux was added to facilitate copper smelting?

They should have wondered why it is indeed intriguing that during the construction of the Tiwanaku and Pumapunku monuments around AD 500, there was a remarkable revolution in ceramic technology. American anthropologist John Wayne Janusek sheds some light on this in his book "Ancient Tiwanaku" (2008, page 22), stating: "(…) Tiwanaku 1, dating from AD 500 to 800, begins with the sudden appearance of a new range of elaborate, decorated red-slipped (redware) pottery. In a significant departure from the Late Formative period, nearly everyone now had access to intricate ceramic vessels for everyday use and festivities."

Janusek's work suggests that this innovation in ceramic technology may have involved the introduction of new chemical elements to the ceramic paste, a type of terracotta fired at relatively low temperatures (around 600 °C) and does not require glazing. To the best of my knowledge, the most effective way to enhance its characteristics is by doping the paste with chemical elements similar to those used in modern geopolymer reactions. My book, "Geopolymer Chemistry and Applications" (Davidovits 2020), describes this technology called L.T.G.S. (Low-Temperature Geopolymeric Setting of Ceramics). Ralph presented and discussed this concept during his teaching and lecture at UCSP in Arequipa, as mentioned earlier in this chapter.

At this point, it is essential to note that this is merely an idea or supposition on my part, as we currently lack sufficient research in this specific direction. However, archaeological evidence does indicate that this innovation did indeed occur around AD 500. It enabled the "industrial" production of pottery, reducing fuel consumption while simultaneously enhancing mechanical properties, shock resistance, fire resistance, and not to mention the decorative aspect.

In essence, the innovation in ceramic technology during the construction of the Tiwanaku and Pumapunku monuments is a remarkable example of Homo sapiens' inventive spirit. This ability to evolve can be easily applied to other activities involving minerals, such as the development of a recipe for creating "malleable" stone, like geopolymer stone. However, it requires placing trust in the intellectual capabilities of Homo sapiens.

Curiously, when it comes to building with stone, archaeologists and anthropologists seem to believe that the Homo sapiens living around AD 500 were incapable of achieving similar technological feats. They seem hesitant or unwilling to move beyond the methods of the Neolithic era, relying solely on abrasive sand and stone hammers as their supposed "scientific proof."

One must wonder why academics did not form an interdisciplinary research team to investigate the legends their people shared regarding the existence of "malleable stone" (as discussed in Chap. 1). Not only would this have been a significant scientific contribution, but also a testament to their principles. Unfortunately, they chose to remain stagnant in the Stone Age, closing the door to what experts, sculptors, stonemasons, and architects have identified as a genuine possibility. It is no wonder that the extraterrestrial theory has gained popularity in the public's imagination.

Before we proceed to Pumapunku, it is crucial to delve into its history. A start in history is my primary approach when studying archaeological sites. When it comes to Andean civilizations, it raises a recurring question: How do archaeologists and anthropologists establish a chronology in the absence of written materials? I raise this question because the Altiplano cultures had none, unlike other Amerindian civilizations like Mexico and Guatemala, which have hieroglyphic writing. No texts, no engraved tombstones, no papyrus, nothing.

Faced with such silence, how can one attempt to reconstruct a comprehensive history? This is a question that continues to challenge me.

References

Davidovits J (2020) Geopolymers in ceramic processing. In: Geopolymer chemistry and applications, 5th edn. Geopolymer Institute, Saint-Quentin (France), pp 563–588

Fernandez JM, Duran S, Yanarico GK (2017) Un aporte a la Arqueología Experimental mediante los procesos de manufactura en la elaboración de la réplica de la Estela el "Fraile" de Tiahuanaco. In: Paper presented at the Vi Simposio Latinoamericano de Física y Química en Arqueología, Arte y Conservación del Patrimonio Cultural, LASMAC 2017, La Paz Bolivia, 10–14 July 2017
Websites: www.dramatic.fr, www.unsacsurledos.com

Chapter 4
Understanding the History of Tiwanaku/Pumapunku

Let us return to the situation in 1550, according to Cieza de León—The high-tech ceramic apogee at Tiwanaku used as a dating method—The rise of the water level of Lake Titicaca at the origin of the brilliant civilization of Tiwanaku—The technological revolution in hand with a period of peace.

Indeed, the absence of written records from the ancient civilizations of South America, particularly the Altiplano region, presents a unique challenge for researchers and historians. While civilizations in the Middle East and Europe have left behind texts that allow us to explore thousands of years into history, South America had to rely on the accounts of early Spanish explorers and chroniclers.

One such important figure in our pursuit of knowledge about pre-Columbian history is Pedro Cieza de León [1520–1554]. Cieza de León, an esteemed chronicler and historian of the Andean world, published his renowned four-volume work, Crónica del Perú (Chronicle of Peru), in Seville, Spain, in 1553. In the first volume, he vividly describes his visit to Tiwanaku and Pumapunku between 1549 and 1550. His detailed account provides us with a wealth of valuable information, shedding light on the historical phenomenon of this civilization that lacks written or engraved material, a complexity that Cieza himself openly acknowledges.

It is important to note Cieza de León's significance in the field. Arriving in America at the young age of 13, he spent nearly two decades participating in numerous expeditions and witnessing the establishment of various cities. His Chronicle of Peru stands as a pioneering project in providing historical testimony on the Andean world, earning him the title of "the prince of the Spanish chroniclers."

With Cieza de León's invaluable accounts and observations, we can investigate the mysteries of the past and better understand the rich history of the Andean civilizations.

J. Davidovits, *Ancient Geopolymers in South America and Easter Island*,
SpringerBriefs in Earth Sciences, https://doi.org/10.1007/978-3-031-75336-7_4

4.1   Let Us Return to the Situation in 1550, According to Cieza de León

We will analyze and transcribe here the whole of Chapter 105 of Crónica del Perú (Chronicle of Peru), which I have arbitrarily divided into four parts according to a thematic classification to facilitate the reading and analysis of the text. Thus, we will be covering the following topics:

(1) Description of the Tiwanaku temple in the village of "Tiaguanaco."
(2) Data on Pumapunku.
(3) The Incas: general comments on the situation and history of this region of Peru.
(4) No written work: Reflections on the need for written records and the complexity of interpreting history in the absence of such testimonies.

Translated from Spanish into English.

CHAPTER CV.

Description of the Tiwanaku temple in the village of "Tiaguanaco."

(1) *Tiaguanaco* may not be a large town, but its notable feature lies in the impressive and remarkable buildings it possesses. Adjacent to the main structures is a man-made hill constructed upon substantial stone foundations. Beyond this hill, you will encounter two exquisitely carved stone idols in the likeness of humans, displaying remarkable craftsmanship and well-defined features. These statues, crafted with such expertise, give the impression that they were the work of skilled artisans or masters. Their size is so immense that they resemble small giants, and their attire distinctly differs from that of the natives of these provinces. It appears that their heads bear some form of ornamentation. In close proximity to these stone figures stands another structure, ancient and devoid of inscriptions. It is this absence of written records that prevents us from identifying the people who constructed such magnificent foundations and walls, as well as the time that has passed since their creation. What remains now is a remarkably constructed wall, a testament to its age, with certain stones showing signs of wear and erosion. In this area, several stones of enormous size exist, leaving one astounded at the sheer magnitude and questioning how human strength alone could have moved them to their current position. Many of these stones are intricately carved, some even taking on the form of human bodies, likely meant to represent their idols. Alongside the wall, numerous cavities and hollows lie beneath the earth's surface.

However, as depicted in Fig. 4.1, an illustration created by the French diplomat Léonce Angrand in 1848, we unfortunately cannot fully visualize the state of Tiwanaku during Cieza de León's visit in 1553. In this drawing, made nearly three centuries later, Angrand portrays a series of isolated blocks arranged as a rectangular alignment of "menhirs," with the Gate of the Sun appearing fractured and distorted.

Unfortunately, Tiwanaku fell victim to quarrying in 1635, leading to the utilization of its stones for construction purposes in Tiahuanaco village and the surrounding

Fig. 4.1 View of the ruins of Tiwanaku in 1848, according to Léonce Angrand. The alignment of the monoliths is what remains of the enclosure of the temple "Kalasasaya" and the Akapana pyramid called "the hill" in the text. The white arrow points to the broken Sun Gate, and the black arrow points to two persons who give an idea of the dimensions of these red sandstone pillars (Bibliothèque Nationale, Paris)

areas, including the church. While most of the stones were repurposed, the immense red sandstone pillars were too massive to be cut into smaller blocks. Consequently, Leonce Angrand was unable to witness the grand wall described by Cieza de León. As depicted in Fig. 4.2, in the modern restoration efforts, these monoliths now serve as pillars for the temple.

Despite the persistent destruction that many places of worship have endured throughout history, the Puerta del Sol, or Sun Gate, has managed to survive, albeit in a fractured state. This can be observed in Fig. 4.3, which presents a comparison between a photograph from 1876, where the ground is littered with debris, and the gate is partially buried, and another photograph showcasing its condition in 2017.

Fig. 4.2 View of the restored Kalasasaya, taken from the exact location as in Fig. 4.1

Fig. 4.3 The Puerta del Sol (Sun Gate). **a** On the left photo is Alphons Stübel in 1876, **b** and on the right is the current restoration

Despite its imperfections, the Puerta del Sol has endured the test of time, offering us a glimpse into the past.

(2) *Data on Pumapunku:* Cieza de León continues his account by describing the remarkable ancient remains in another area, further west of the main building—Pumapunku. He notes the presence of numerous doorways, each consisting of a single stone encompassing the jambs, lintels, and thresholds. As he observed and recorded these details, what struck him the most was the larger stones upon which the doorways were constructed. Some of these stones measured an impressive thirty feet in width, fifteen feet or more in length, and six feet in thickness. The unity of these grand doorways, carved from a single stone, exuded a sense of splendor and magnificence when carefully examined. Cieza de León confessed his bewilderment regarding the tools and instruments that could have been used to carve such monumental stones. It is evident that these stones must have been even larger before being crafted and perfected. The unfinished nature of these structures becomes apparent, as only these portals and other remarkably large stones are present. Cieza de León observed some of these stones already shaped and dressed, ready to be placed in the building, which stood slightly to one side. Additionally, he mentions the existence of a great stone idol, presumed to be worshipped, with rumors circulating of the discovery of gold nearby. Surrounding the temple, Cieza de León noted several other large and small stones, all expertly dressed and carved, similar to those he previously described.

Reading Cieza de León's account, one can easily imagine his sense of awe as he contemplated the immense red sandstone megaliths that formed the platform and the volcanic andesite doorways made from a single piece of stone. "...which is one of grandeur and magnificence when well considered..." he pondered, unable to comprehend the methods or tools used to achieve such monumental stonework. It is safe to assume that Cieza de León found Pumapunku even more captivating than Tiwanaku.

By 1550, the gold leaf that once adorned many of the gates had vanished, a structural characteristic that we will explore further later on, as it is one of the outcomes of the andesite-made geopolymer stone production method.

A century later, during what we refer to as the "destruction" phase, the colossal red sandstone platforms proved too challenging to be cut into smaller sizes for use as building materials. The same fate befell the gates, which were knocked down and left in broken fragments. On the other hand, the H-shaped carved structural elements, presumably made from a volcanic rock similar to andesite, presented their own set of difficulties. These elements were incredibly hard, making them nearly impossible to cut. Moreover, their shape did not lend itself to being reused as bricks or building blocks, suggesting they were likely destroyed through a forceful impact. During their visit, our team selected a piece of this particular stone type, which will be discussed in detail in Chap. 8.

The geopolymer andesite stone used in these structures possesses a remarkable level of compactness and durability. Its toughness is comparable to that of the hardest natural stones, often employed for making grinding stones, for instance. As a testament to its strength, some slabs of this material were later repurposed as grinding stones, leaving behind a visible trace in the form of a millstone next to a stack of stone bricks at Tiwanaku (see Fig. 4.4).

(3) *The Incas; general comments on the situation and history of this region of Peru*: Cieza de León provides intriguing insights about the Incas and their connection to Tiwanaku. He concludes that Tiwanaku is the most ancient site in all of Bolivia, even predating the time of the Incas. According to local indigenous inhabitants, the Incas built their impressive structures in Cuzco based on the architectural plans they observed in Tiwanaku. In fact, it is believed that the

Fig. 4.4 A grinding wheel from a andesite geopolymer slab. Tiwanaku, near the Kalasasaya

first Inca rulers even established their court and seat of power in Tiwanaku. One striking aspect is the absence of nearby rock quarries in the region, which raises questions about the source of the countless stones used in construction. The transportation of these massive stones would have required a substantial workforce. When he asked the natives, in the presence of Juan de Vargas, who oversees them through an encomienda [labor system involving a community of natives], whether these structures were built during the time of the Incas, they chuckled at the notion. They confirmed what he had already suspected: these structures were erected long before the Incas came to power. However, they could not provide any specific detail about who built them, except for the oral tradition passed down from their ancestors, which suggests that it was accomplished in a single night. These stories, coupled with their tales of bearded men on the island of Titicaca and the construction of Vinaque by such individuals, lead him to speculate that prior to the Inca reign, there may have been a group of intelligent individuals who arrived from an unknown place and accomplished these extraordinary feats. Perhaps, being vastly outnumbered by the native population, these people were eventually wiped out in wars and conflicts.

Indeed, Cieza de León's inquiries about the builders of Tiwanaku/Pumapunku to the local Indians, whom he refers to as the "Ingas," are met with laughter. From their perspective, the idea that the Incas, seen as invaders like the Spaniards, constructed these ancient monuments seems absurd. The Indigenous, loyal to their ancestral heritage, refuse to disclose the secrets surrounding the works of their distant predecessors. However, they do provide a tantalizing clue—the monument's construction occurred in a single night. The means by which this was accomplished remains a mystery.

It is important to note that Cieza de León communicates with the inhabitants through interpreters, and the responses he receives are in the Aymara language, which is then translated into Spanish via Quechua. Translating and interpreting technical language, such as gestures or discussions of technological knowledge, can indeed be complex. In the case of Cieza de León's text, it is unclear whether the reference to the construction taking place in one night pertains to the entire monument or specifically to the manufacture of the artificial geopolymer stone, which may be the true secret. Considering the popular tradition surrounding the technology of creating "malleable" stone, it is plausible to propose the latter hypothesis, suggesting that the geopolymer stone, whether red sandstone or gray andesite, hardened overnight. It is worth noting that we have the capability to replicate this knowledge in laboratory settings.

(4) *No written work*: Reflections on the need for written records and the complexity of interpreting history in the absence of such testimonies. The lack of written records in the New World has left us in a state of ignorance regarding many aspects of its history. Cieza de León acknowledges the tremendous value of the invention of writing, which allows the memory of events to endure for centuries and ensures their fame spreads far and wide. Thanks to the availability of written

texts, we are not left in the dark; we have the privilege of accessing knowledge at our fingertips.

However, no such written records have been discovered in this new world of the Indies. Consequently, countless aspects remain shrouded in mystery. In addition to the remarkable structures discussed by Cieza de León, there are also the residences of the Incas and the very house where Mango Inga, son of Guaynacapa, was born. Adjacent to these dwellings stand two imposing tombs resembling towering structures with wide corners and doors facing the rising sun. These tombs are believed to belong to the native chiefs of this town.

The lack of written accounts hampers our understanding of the significance and stories behind these structures and their inhabitants. We can only imagine the tales that could be unveiled if we had access to the written records that have been absent thus far.

Cieza de León's mention of the house where Mango Inga, son of Guaynacapa, was born refers to the renowned Inca emperor Manco Capac, who was the son of Huayna Capac, also known as Wayna Kapac in Quechua. Manco Capac played a crucial role in the history of indigenous resistance against the Spanish conquistadors. He organized and led the rebellion against the invaders until his assassination in 1544, several years before Cieza de León visited the region.

This raises an intriguing question posed by American anthropologist John Wayne Janusek: Did the Inca rulers merely claim ancestral ties with the ancient kings of Tiwanaku to legitimize their own reign? Did they declare Tiwanaku a sacred site of cosmic origin and seek to emulate its impeccably crafted monuments to solidify their rule? The question of who actually created Tiwanaku remains unanswered, shrouded in mystery and speculation.

On page 104, Janusek (2008) notes that: "(...) Between AD 500 and 600, Tiwanaku crashed the party as the most prestigious centre in the Lake Titicaca basin and the south-central Andes. The question, "Why Tiwanaku?" as yet yields obscure answers (...). However, there was no preordained reason for this settlement to become an unparalleled cultural, economic and political centre. Tiwanaku commanded no inherently privileged landscape or important resource (...)".

4.2 The High-Tech Ceramic Apogee at Tiwanaku Used as a Dating Method

Upon revisiting the statements made by the anthropologist, I respectfully disagree with his conclusions. It seems that, like some of his young colleagues at the LASMAC 2017 conference, he may have overlooked the significance of a crucial technology highlighted in the realm of ceramic craftsmanship, as detailed in Chap. 3 of his work. I believe it is important to reiterate this point, as it provides valuable insight into the Tiwanaku culture.

In Chap. 3, the anthropologist describes the emergence of a new style of pottery known as "redware" during the Tiwanaku 1 period, dating from AD 500 to 800. This marked a significant departure from the Late Formative period that preceded it. Suddenly, there was widespread availability of intricately decorated red-slipped pottery, allowing almost everyone to possess exquisite ceramic vessels for everyday use and ceremonial feasting. This shift in ceramic production and consumption patterns represents a notable development in Tiwanaku society.

By emphasizing the transformative impact of the Tiwanaku ceramic "boom," we gain a deeper understanding of the time's cultural dynamics and technological advancements (Janusek et al. 2013). It is crucial to acknowledge and consider the significance of such findings when examining the broader narrative of the Tiwanaku civilization.

4.2.1 Ceramics as a Dating Method

The question of determining historical periods in the absence of texts or engraved stelae has also intrigued me, especially when studying South American civilizations. Fortunately, anthropologists recognized early on that the study of ceramic art, pottery, and clay work could provide valuable insights.

The unique climatic conditions of the Altiplano, with its high altitudes exceeding 3800 m above sea level, have a significant impact on the environment and its inhabitants. The intense ultraviolet (UV) rays, combined with the high humidity, present challenges for preservation. These challenges are evident in the degradation of textiles, making it difficult to obtain relevant archaeological information from them. Textiles require specific conditions for their preservation to reveal their richness to archaeologists.

In this context, ceramic and stone emerge as the materials that withstand the relentless erosion caused by the climate. They become invaluable resources for studying and understanding the culture. Prior to the discovery of techniques like carbon-14 or thermoluminescence dating, the evolution of ceramics was the primary method used by anthropologists to make progress in their understanding of the civilization.

It is important to note that this method relies on the comparative analysis of shapes, designs, and paintings, interpreted through an artistic lens rather than relying solely on scientific data. This approach formed the basis for establishing the chronology, which I will develop further at the end of this chapter.

4.2.2 The Role of Tiwanaku High-Tech Pottery

Figure 4.5 showcases three distinct examples of ceramics crafted in Tiwanaku. On the left, we have a piece dating back to an earlier period, while on the right, we see the iconic red ceramic vessel, also known as "redware," which signifies the ceramic

Fig. 4.5 a On the left, an Early Period vessel (AD 200); **b** in the center, the famous kero; **c** on the right, the tazon, a multicolored red vessel (AD 500) (Museo Nacional de Arqueología Tiwanaku, La Paz)

revolution I previously mentioned. The central piece in Fig. 4.5 is the renowned black and red vessel called the "kero," and on the right, the "tazon," which enjoyed fame throughout various regions of the Altiplano. These vessels were meticulously made from a fine and solid ceramic paste, in stark contrast to the thicker composition of the older pottery.

I believe achieving such technical excellence in ceramics necessitates a remarkable process involving a geopolymer-type chemical reaction within the clay's constituent elements, particularly kaolinite or ferro-kaolinite. This technique falls under the "LGTS" technology category, which stands for Low-Temperature Geopolymeric Setting of Ceramics. It enables the production of robust ceramics at temperatures ranging from 400 to 600 °C, aligning with the fuel availability and quality of fireplaces used in the Altiplano. Historical records indicate that the primary combustibles were camelid dung (from llamas, alpacas, etc.), grasses, and shrubs from the savannah. Under normal circumstances, these conditions would yield common "terracotta" pottery fired at temperatures lower than 600 °C.

A recent study conducted by Sidoroff (2019) replicated the firing conditions using camelid (dromedary) dung in the arid desert of Jordan. In ten firing attempts, her team achieved an average maximum temperature of 596 °C. These temperatures, albeit relatively low, were achieved using open fires, similar to garden bonfires. Higher temperatures would necessitate enclosed kilns, which, based on archaeological findings, do not appear to have existed in Tiwanaku and its surrounding region. However, smaller fireplaces may have been used for lime production, requiring temperatures of at least 650 °C.

For more detailed information on LTGS geopolymer technology and its applications in archaeology, the interested reader may find relevant publications and online resources through the provided links at the end of this book. The LTGS technique encompasses two main approaches:

(1) The addition of an alkaline reagent, typically natron salt (sodium carbonate) mixed with lime and water, to generate caustic soda (sodium hydroxide) NaOH, resulting in geopolymerization in an alkaline environment or;

(2) The addition of an acid reagent, often phosphoric acid, derived from the reaction between an organic acid extracted from plants (such as acetic acid/vinegar,

lactic acid, citric acid) and guano (calcium phosphate and oxalate) or bone powder (calcium phosphate hydrate), facilitating geopolymerization in an acidic environment.

These reagents are usually introduced during the clay's maturation phase. For now, it suffices to acknowledge the existence of these geopolymer chemical reactions.

Tiwanaku's "high-tech" pottery, with its unique characteristics, holds immense significance and will serve as a reliable dating method in several chapters of this book. In Chap. 7, we will explore how this technique sheds light on the volcanic materials employed in the geopolymer andesite rock used to construct the magnificent gates and H-shaped structures of Pumapunku.

4.3   The Rise of the Water Level of Lake Titicaca at the Origin of the Brilliant Civilization of Tiwanaku

Another crucial aspect to consider in developing and evolving pottery production is the indispensable requirement of abundant water during its primary stages. However, being situated at an elevation of 3800 m, the Altiplano experiences low rainfall. The primary water source in this region is the Tiwanaku River, which nourishes Lake Titicaca. If the Tiwanaku/Huaquira River and its tributaries were to dry up, it would lead to a significant drop in the lake's water level, resulting in an ecological catastrophe. Historical records show that the Altiplano has witnessed several of these events in the past, with severe impacts on civilizations and life surrounding Lake Titicaca. In fact, the cyclical pattern of dry and wet periods is the underlying cause behind the rise and fall of successive civilizations in the area.

During the 1990s, paleo-ecological studies shed light on the abrupt climate changes, the influence of phenomena like El Niño, and the alternating dry and wet periods that caused noticeable variations in Lake Titicaca's water level. Scientific teams from North America and Europe utilized two primary techniques to investigate these phenomena: the analysis of lake sediments and the study of the Quelccaya glacier. Additionally, carbon-14 dating was employed on these sediments.

The findings of these studies have been synthesized in a graph, presented in Fig. 4.6, which illustrates the long-term fluctuation of Lake Titicaca's water level from 2000 BC to AD 2000. The thickness of the curve indicates the seasonal variations in the lake level, which typically average around 5 m. The height of the water level is measured in relation to the bottom of the Desaguadero River, which is considered the "0" level. Currently, the water height at the river estuary ranges from 5 to 6 m. When the level falls below "0," the river water flow becomes minimal or may even completely dry up. Another river that experiences reduced flow or drought is the Tiwanaku/Huaquira River. In such scenarios dominated by these events, the lake's shores recede by several kilometers.

The graph depicted in Fig. 4.6 vividly illustrates the dramatic fluctuations in water levels, including prolonged periods of severe drought in the region. One notable dry

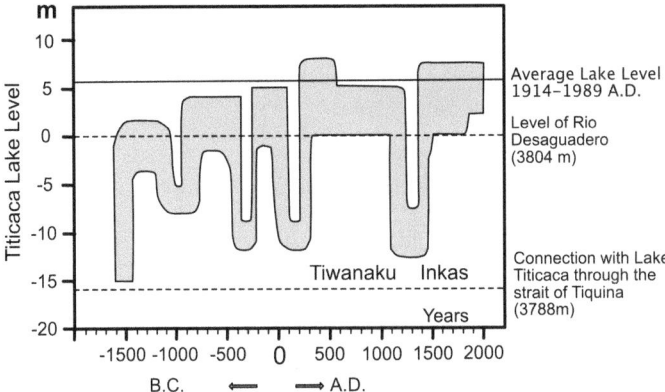

Fig. 4.6 Changes in the level of Lake Titicaca between 2000 BC and 2000 AD. After Abbott et al. (1997)

spell occurred just prior to the rise of the Inca Empire, between AD 1100 and 1300. However, the most significant event for our archaeological research occurred around AD 100–300 when the lake level experienced a sudden drop of 12–15 m. As a result, the lake's shores receded by 10–15 km, bringing about a drastic change in the local living conditions.

Following this period, there was a sudden increase in the water level after AD 300, with Lake Titicaca reaching a height of 7 m. This indicates that the Tiwanaku River, along with other rivers, had resumed its flow, providing pottery craftsmen with favorable conditions to practice their art. The clay available was of good quality, and water was abundant. It was during this time that the "alchemical" innovation, involving the addition of specific chemicals and organic ingredients like plant extracts, took place. These circumstances created the ideal context for the emergence of what we can consider as "high-tech" pottery, surpassing the pottery produced in other cities of the region.

4.4 The Technological Revolution in Hand with a Period of Peace

Upon analyzing the data, it becomes evident that a technological revolution took place during this period, serving as the foundation for the flourishing Tiwanaku civilization and its religious practices. This growth fueled extensive trade networks, with countless llama caravans traversing the Altiplano from north to south and east to west. It also provides evidence that the knowledge required for the creation of artificial stones, such as the "red sandstone" used in megalithic terraces and the "andesite" utilized in sculptures and "H" gates, not only existed but was abundantly

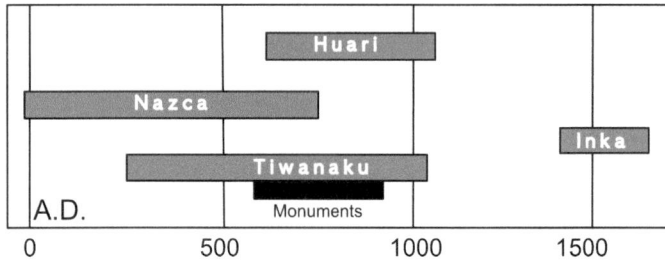

Fig. 4.7 Chronology of the three contemporary civilizations of the Andes: Nazca, Tiwanaku, Wari (Huari), and that of the Incas (Inka). The period of construction and use of the two sites of Tiwanaku and Pumapunku before their destruction is indicated by the word "monuments"

available. This offers an initial response to the question we posed at the beginning of this chapter: Why Tiwanaku?

Tiwanaku did not emerge out of thin air or as a result of extraterrestrial influence. The artisans of Tiwanaku and Pumapunku possessed and practiced this knowledge and expertise. The accumulated experience in LTGS technology and pottery production allowed them to construct colossal structures that we will explore in the subsequent chapters. I dare to propose that this technological revolution can be compared, for now, to the "Digital Revolution" of the twenty-first century. The Tiwanaku Valley of the sixth-seventh century was, in a sense, our modern-day equivalent of "Silicon Valley."

It is worth noting that the abrupt climate changes described also impacted other regions of the Andes. The Nazca civilization, renowned for its gigantic geoglyphs, and the Huari (Wari) civilization, which laid the foundation for Cuzco, are both widely recognized, surpassing the fame of Tiwanaku. As depicted in Fig. 4.7, all three civilizations existed concurrently.

What is truly remarkable is the peaceful coexistence, good neighborly relations, and remarkable tolerance exhibited among these three cultures and geographical entities. They managed to connect and interact without significant conflicts, wars, occupations, or annexations. A prime example of this is the harmonious territorial coexistence between Tiwanaku and Huari (Wari), as depicted in Fig. 4.8. In the region between Arequipa and Moquegua, specifically in the Moquegua Valley, Tiwanaku settlements were established around AD 400. In contrast, Huari (Wari) settlements emerged just a few kilometers away between AD 600 and 650.

As mentioned in Chap. 2, the OMO temple, administered by Tiwanaku, finds its counterpart in the Huari (Wari) civilization's Cerro-Baúl. This situation is highly unusual and possibly unique in human history. It becomes even more intriguing when we consider that a few centuries later, following the drought period between AD 1100 and 1300, the region witnessed the rise of the Inca Empire through force. This subject has been extensively studied and theorized.

After these reflections and preambles, the time has come to embark on our exploration of Tiahuanaco. Let us investigate its wonders and unravel its mysteries.

Fig. 4.8 Zones of shared (hatched area) influence between Tiwanaku and Huari (Wari)

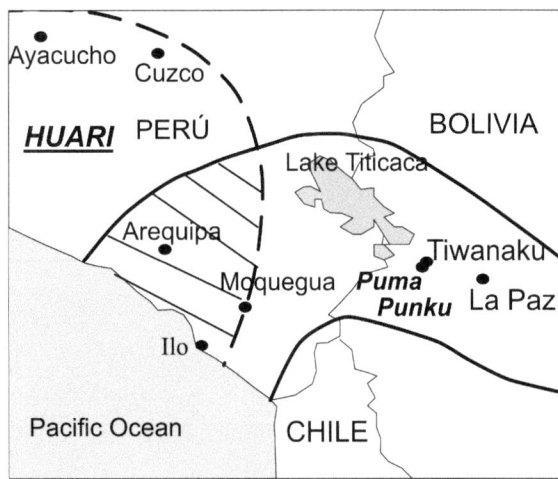

References

Abbott MB, Binford MW, Brenner M, Kelts KR (1997) A 3500 C14 year high-resolution record of water-level changes in lake Titicaca, Bolivia/Peru. Quat Res 47:169–180. Article No. QR971881

Janusek JW (2008) Ancient Tiwanaku. Cambridge University Press, New York. ISBN 978-0-521-01662-9

Janusek JW, Williams PR, Golitko M, Lémuz AC (2013) Building taypikala: telluric transformations in the lithic production of Tiwanaku. In: Tripcevich N, Vaughn KJ (eds) Mining and quarrying in the ancient andes. Interdisciplinary contributions to archaeology. Springer Science+Business Media, New York, pp 65–97

Sidoroff ML (2019) Experimental bonfirings of pottery with camel dung fuel, Jordan, July 2018. https://exarc.net/ark:/88735/10427. Issue 2019/2

Chapter 5
First Day in Tiwanaku/Pumapunku

The first visit to the archaeological sites: Pumapunku/Tiwanaku—The team heads to Pumapunku—The Four Megalithic Red Sandstone Terraces—Geometric sculptures in the grey "andesite volcanic rock"—Tiwanaku—The Puerta del Sol (Sun Gate)—Summary of the first visit to Pumapunku/Tiwanaku.

In the afternoon of Wednesday, November 9, 2017, Ralph and Luis are dropped off by their driver just before the border bridge at Desaguadero, which connects Peru and Bolivia (as shown in Fig. 5.1).

The driver is unable to proceed any further as the bridge is exclusively for pedestrians. This location holds significance as it is precisely on the banks of the Desaguadero River where scientists determined the "0" level while analyzing the water level variations in Lake Titicaca, a study mentioned in the previous chapter. After crossing the bridge, Ralph and Luis go through the customary customs and police procedures, which turn out to be quite efficient, unlike the long queues often experienced by travelers and tourists. With the entry formalities completed, they take a taxi to their accommodation in the capital city, La Paz, opting for a district near the airport.

5.1   The First Visit to the Archaeological Sites: Pumapunku/Tiwanaku

While I had previously mentioned it, I believe it is worth reiterating the unwritten rule that has been established when it comes to naming these archaeological sites: Tiahuanaco is the Spanish name for the village, while Tiwanaku is the name archaeologists use to refer to the archaeological monuments and the history of this civilization. As for Pumapunku, although it can be found written as two words, the standard usage seems to favor using it as one word, as Carlos Ponce Sangines indicates in the title

J. Davidovits, *Ancient Geopolymers in South America and Easter Island*, SpringerBriefs in Earth Sciences, https://doi.org/10.1007/978-3-031-75336-7_5

Fig. 5.1 Desaguadero, border point on the river, on the left Peru, on the right Bolivia; Ralph on the bridge with Lake Titicaca in the background

of the report mentioned in Chap. 2 and as confirmed by the descriptive panel at the entrance of the site itself.

Following my study protocols for archaeological sites, our team's visit began with a tourist excursion, which provided us with a general overview before we went into a more detailed survey. On Thursday, November 10, the field team went on a minibus trip to Tiahuanaco. Throughout the day, they explored the sites of Pumapunku and Tiwanaku, capturing the entire experience through photographs that I would receive later that day. As depicted in the satellite photo shown in Fig. 5.2, Pumapunku is situated just a few hundred meters from Tiwanaku, on the other side of the road. They appear to exhibit a dualistic organizational structure, with each population maintaining its distinct customs and traditional attire.

In his book, as mentioned in Chap. 2, Carlos Ponce Sangines sheds light on this situation on page 25, and I quote:

> It seems reasonable to assume that in the pre-Columbian city of Tiwanaku, a dualistic organization was already established, with a system of divisions in the north and south. Speculatively speaking, it can be suggested that the temple assigned to the north was Kalasasaya [Tiwanaku], and the one assigned to the south was Pumapunku. (...) It is important to reiterate that this type of division does not imply social differentiation based on economic dominance or racial discrimination. To avoid any confusion, let us recall the definition advocated in the introductory paragraph.

A similar situation of territorial coexistence, as mentioned in Chap. 4, can be observed in Moquegua between the people of Tiwanaku and those of Huari (Wari) (as shown in Fig. 4.8).

As I have emphasized from the beginning, our primary objective is to study the red "sandstone" megaliths. These massive and immovable structures suggest that

Fig. 5.2 General view of Tiahuanaco on Google Earth; top right, Tiwanaku. Bottom left, Pumapunku

they were constructed using artificial stone. Determining their origin is our precise aim. Of course, we will also study other artifacts present, but they do not hold the same level of priority.

Why do we prioritize the study of the red "sandstone" megaliths? It is because we have access to a geological and mineralogical study conducted by Ponce Sangines' team in 1970, which serves as our starting point. Our goal now is to verify the geological sources on-site, utilizing modern analysis methods, and compare the red sandstone found in the geological deposits with the material of the terraces at the archaeological site. If we discover it is a man-made rock, it should exhibit distinct characteristics.

5.2 The Team Heads to Pumapunku

From the roadside, only a few scattered stones are visible, and as the team follows the path towards the site, they come across blocks lying on the ground.

5.2.1 First Impression

Figure 5.3 clearly portrays a landscape that resembles a post-earthquake zone, with a slightly ascending path leading to a platform. As our team researchers make their way up the ascending path, on the left side, they come across a truly remarkable sight—a towering rocky gate apparently made by andesitic volcanics, exquisitely carved and lying on the ground. This gate, when standing, reached an impressive height of nearly 3 m. Its intricate craftsmanship is a testament to the skill of its creators. Just behind the gate, they spot two of the renowned "H" blocks (which I have mentioned in previous instances and can be seen in Figs. 3.2 and 3.4 of Chap. 3).

As they continue their ascent, they reach a horizontal platform, which, according to the explanation provided on the entrance panel, corresponds to the third level of a pyramidal structure that stands five meters high. Upon reaching this platform, Ralph climbs a small mound on the right and captures a photograph of the scene (Fig. 5.4).

Luis, the geologist, stands near point (4) on the left, giving an idea of the site's size. Ralph and Luis are the only visitors at Pumapunku, and as Luis explores the site to gain an overview of its geology, he points out to Ralph that the color of the sandstone megaliths strikes him. He has never encountered sandstone with such a rich, solid shade of red. In this region, sandstone tones tend to be paler, almost pastel, with hints of pale pink or brown. This red color is evidence that this material is indeed man-made since the builders likely incorporated red clay into the geopolymer mixture. This realization confirms that we are on the right track, and tomorrow and the following day, we will examine the colors of the geological sandstones and clays on site.

That evening, after closely examining the photographs sent by Ralph, a question immediately arises: why does Pumapunku exude such an unremarkable and desolate

Fig. 5.3 The slightly ascending path leading to the platform of Pumapunku and its "H" blocks

Fig. 5.4 The horizontal platform of Pumapunku seen from north to south. Numbers (1)–(4) mark the four red "sandstone" megaliths, and numbers (5)–(7) mark some rocky blocks of apparent andesitic composition

atmosphere? Both Luis and Ralph share the same impression. In this context, I cannot help but recall Ponce Sangines' statement from 1970, which I mentioned in Chap. 2: *"(...) In this sense, the present work, on the one hand, tends to revalue and draw attention to Pumapunku, which when restored will occupy a pre-eminent place not only in Bolivia's monumental heritage but also in South America's (...)."*

It suggests that Pumapunku's fame was meant to surpass that of Tiwanaku, and this sentiment is further reinforced when reading the text written on the welcoming panel at the site (Fig. 5.7): *"Its name in Aymara means the Gate of Puma. It was the most magnificent structure at Tiwanaku, showcasing the immense proportions of the constructions and the impeccable craftsmanship of its platforms, portals, and lintels."*

5.2.2 The Gates in Grey "Andesite Volcanic Rock"

It is evident that the Gate of the Sun is a major attraction for tourists visiting Tiwanaku. However, it is quite incongruous that Pumapunku, the site within the vicinity, boasts at least three similar gates. The dimensions of these gates can be found in Table 5.1, based on measurements published by Stübel and Uhle in 1892.

These gates, which astonished Cieza de León (as mentioned in Chap. 4), were described in his text, particularly in part (2): *"In another place further to the west of this building are other greater ancient remains, among them many doorways with*

Table 5.1 Dimensions of the various gates found at Pumapunku and Tiwanaku after Stübel and Uhle (1892)

	Puerta del Sol Sun Gate (m)	3 gates at Pumapunku (m)	Puerta de la Luna (m)
Total height	> 3	2.50 à 2.80	2.23
Total width	3.80	2.80 à 3.0	1.80
Width of the jambs	1.45	1.20 à 1.30	0.50
Entrance height	1.90	1.90	–

their jambs, lintels, and thresholds, all of one stone...". Interestingly, in part (1) of the text, the narrator did not mention the Sun Gate at Tiwanaku.

These three grand gates were neatly chiseled, but unfortunately, they now lie broken, cracked, and scattered on the ground, hidden from view. Originally, these gates were crafted from a single monolithic block. In Fig. 5.3, you can see one of these gates lying on the left, just before the entrance, at the base of the two "H" blocks. However, in my opinion, the most significant of these gateways is located behind the remains of megalith No. 3, marked as number (8) in Fig. 5.5.

Additionally, Fig. 5.6 clearly displays that two-thirds of the lintel has been broken, suggesting that it was initially meant to have three rows of longitudinal engravings, similar to those found on the Puerta del Sol, which features four rows. The visible remains of the lintel are severely damaged. We now know that the gates were once covered with gold sheets, a detail that Cieza de León did not mention in his account.

Fig. 5.5 View from south to north of the Pumapunku platform, following the same numbering as shown in Fig. 5.4, and adding the number (8), corresponding to the gateway in andesite volcanic rock

This was the cause of their destruction. While there is a possibility that the conquistadors were responsible for this destruction, archaeologists lean towards a second hypothesis, suggesting that the indigenous people vandalized the site during the general looting of Tiwanaku/Pumapunku, which occurred around AD 800–900.

Interestingly, the Puerta del Sol was spared from destruction precisely because it was not adorned with gold—an aspect we will investigate later. Figure 5.6 shows gate number (8) and its damaged lintel, located next to the remains of megalith Nr. 3 in red "sandstone".

However, since there is no text, how do we know this? The knowledge about the presence of gold decoration on the gates comes from recent studies conducted by Thomas Gara, an American expert in archaeological prospecting equipment. In Gara (2016) conducted a detailed study focusing on the frieze of the gate, as depicted in Fig. 5.6c. The analysis revealed the presence of numerous small holes drilled into the rock, specifically at the bottom of grooves. These holes, totaling approximately 900 across the site, were used to secure the gold leaf decoration using small rivets made of gold. I had the opportunity to collaborate with Gara on this subject, and our joint scientific paper titled *"Considering Certain Lithic Artifacts of Tiahuanaco (Tiwanaku) and Pumapunku (Bolivia) as Geopolymer Constructs"* was recently published (Gara et al. 2020).

Returning to Cieza de León's account, upon his arrival at the site, he was immediately struck by the presence of these three (or four) immense gates skillfully crafted from a single stone block. It was only after passing through these gates that he discovered the large megalithic terraces, which I have numbered as (1), (2), (3), and (4) in Figs. 5.4 and 5.5. Bolivian archaeologists have used Cieza de León's descriptions to propose a schematic diagram of the historic Pumapunku, which I have adapted from the original reconstruction, as shown in Fig. 5.7.

Initially, the entrance to the three-story "pyramid" was situated in the right sector, spanning from west to east. However, today, access is from the left side. The temple itself occupied a relatively small area and featured three doorways, as indicated in the diagram. The temple's design suggests that it was open and oriented towards the east, facing the direction of the sunrise. It is widely agreed upon by various authors, including Cieza de León, that the construction of this site was never completed.

In more recent times, American archaeologist Vranich (2018) proposed a different model, suggesting that the gates were located within the surface of each megalith. Vranich also presents the idea of four gates, which seems more logical given the presence of four red "sandstone" terraces. However, I hold the belief that Vranich's proposed position is not accurate. This assumption is based on the fact that at the base of each gate jamb, there is a cylindrical or conical perforation measuring 10–15 cm in diameter, as clearly depicted in Fig. 5.6. These perforations were evidently intended to accommodate wooden fastening pins. Therefore, the corresponding bored holes would need to be present on the surface of each red sandstone terrace. However, this is not the case, and I will discuss this further when describing each terrace individually. Consequently, the doorways were located at the entrance of each segment, firmly fixed to the prepared ground for that specific purpose.

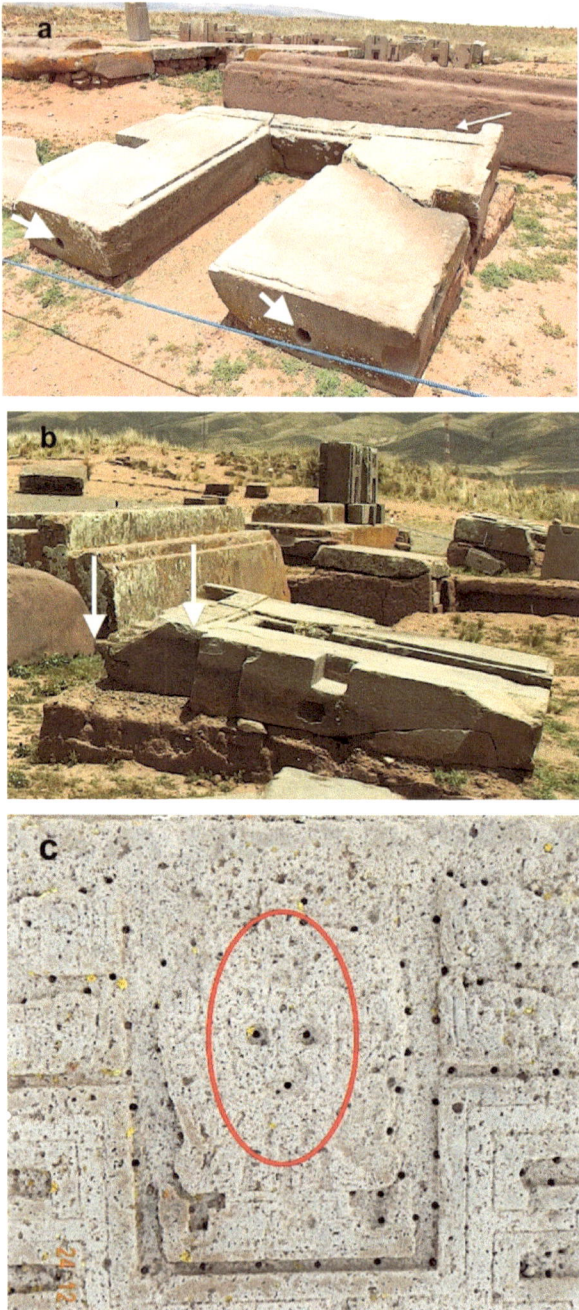

Fig. 5.6 **a** The arrows point to the cylindrical holes dug in each jamb of the gate; **b** the mutilated lintel; the arrows indicate its overall dimension; **c** Detail with the small holes, according to Gara et al. (2020). The red oval indicates the face of a deity

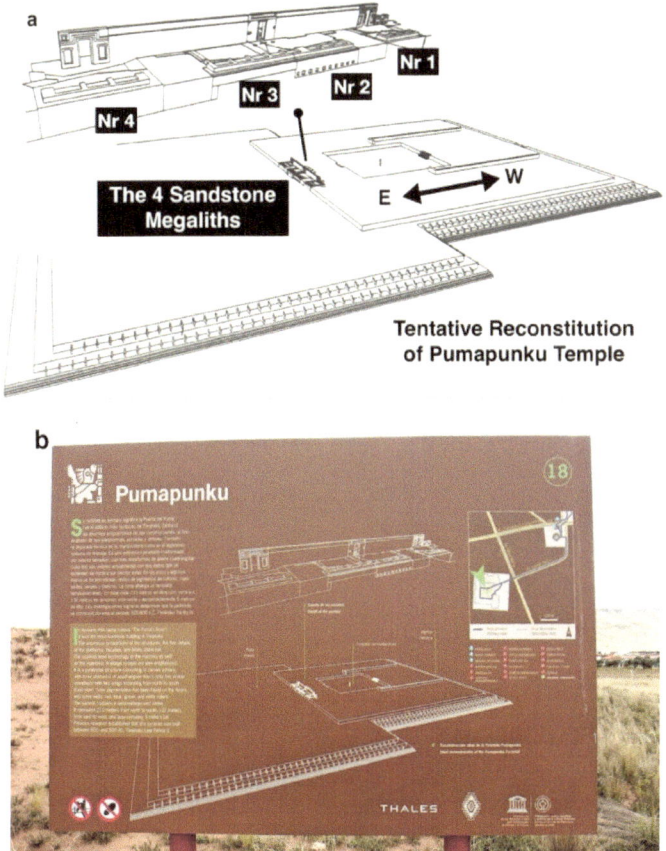

Fig. 5.7 **a** Schematic of the reconstitution of Pumapunku, based on the panel at the entrance; **b** The panel of the historic site

5.2.3 The Four Megalithic Red "Sandstone" Terraces

In Ponce Sangines' 1971 report, he refers to these terraces as "Segment" 1, 2, 3, and 4. Figure 5.8 provides a detailed close-up of these segments. Unfortunately, these terraces have suffered significant damage over time, with many of them broken and divided into multiple parts. Some have even disappeared or remain embedded in the ground. If we consider that these terraces were artificially constructed, then they would ideally be monolithic, similar to how a modern concrete floor is cast in a single piece. However, Segment 2 in Fig. 5.8 may be an exception, and we will explore the reasons behind this shortly.

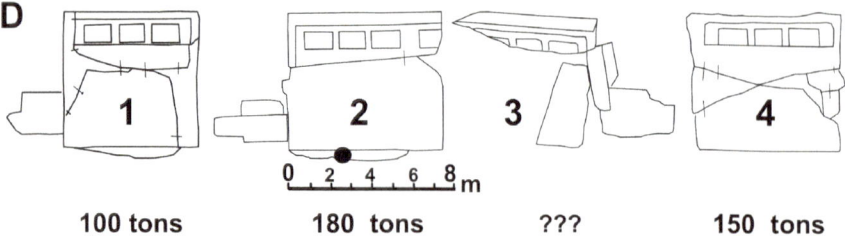

Fig. 5.8 Representation of the four red "sandstone" megalithic segments or terraces with the estimated weight each would have had in its original state

Fig. 5.9 Segment 1; the arrows indicate the location of the cramp sockets into which the metal was cast

On the other hand, if we assume that the terraces were made from natural sedimentary rock, as suggested by various authors, they would have been assembled from several heterogeneous pieces. The precise fitting of these pieces and the use of metal-filled cramp-sockets throughout the structure have, for centuries, fueled speculations about extraterrestrial intervention or the existence of highly advanced civilizations. In the case of hardened artificial monoliths, it is possible that cracks formed due to inconsistencies in the soil's composition or poor curing techniques, potentially exacerbated by factors like earthquakes. Similar to the behavior of modern concrete, fractures can occur if the mixture is poured onto inadequately compacted soil.

Figure 5.9 provides a depiction of Segment No. 1, originally planned to be 7 m long, 6 m wide, and 0.9–1 m thick. Upon examining the surface details, Luis observes that there is no cylindrical cavity within the segment that would accommodate a gate, contradicting A. Vranich's proposal. Similar to the other segments, the end of Segment No. 1 features a bench-like structure with hollow surfaces that resemble seats or armchairs. While some have drawn comparisons to courtrooms or parliamentary chambers, the true purpose of these features remains unknown. Additionally, the cramp sockets, where the metal was cast, can be clearly seen, indicated by the white arrows.

Moving on in Fig. 5.9, we encounter Segment No. 2, located to the right of Luis. This segment is the largest of the four and consists of two parts: a horizontal part and a vertical part. Combined, their original weight was approximately 180 tons. The horizontal part, weighing around 100 tons, spans about 8 m in width, while the vertical part, weighing 80 tons, stands at approximately 1.75 m in height (Fig. 5.10).

The purpose of the ten hollowed-out cavities at the bottom of the vertical part remains a mystery. It is evident that these two parts were likely created separately

Fig. 5.10 Luis in front of Segment 2; **a** the horizontal part and its step; **b** the vertical part on which the 4-seat bench is located

but cast against each other. This approach seems logical, as constructing the assembly in one piece would have subjected the mass to significant stress, resulting in cracks similar to those observed in the other segments constructed as single pieces. In contrast, the separation of the two pieces later on left the horizontal section free from internal tension, effectively preserving its integrity.

Only the bench and a portion of the horizontal plane of Segment No. 3 still remain buried in the ground. Figure 5.11 showcases the lower section of the bench on the left and the remnants of the horizontal floor on the right, as viewed from Segment No. 2. The upper area of the bench has suffered severe erosion, and the imprint of the "seat" has nearly vanished from the perspective of today's visitors. However, it is likely that a century ago, the seat's mark was clearly visible.

Figure 5.12 exhibits Segment No. 4, where a section of the horizontal base connects to the bench, with its three intact "seats". In the center, grass has replaced the stone. Some smaller blocks show less erosion compared to the rest of the segment, suggesting that they may not belong to the original megalith.

Fig. 5.11 a Luis is standing next to the bench of Segment No. 3; **b** the remains of this segment, seen from Segment No. 2

Fig. 5.12 a The assembly of Segment 4; **b** the upper part with the bank. The small blocks do not seem to belong to the original megalith

5.2.4 Geometric Sculptures in the Grey "Andesite Volcanic Rock"

Apart from the gates, which are also crafted from appearing andesite volcanic rock, other sculptures capture the admiration and commentary of visitors. In Figs. 5.5 and 5.6, these sculptures are labeled as (5), (6), and (7). It is important to note that rock andesite has a hardness comparable to quartz (slightly lower), ranking at 6–7 on the Mohs scale, and a density of 2.8 kg/l. These properties make it extremely challenging to cut and carve, an impossibility with stone tools.

Let us begin with the block with small holes labeled as No. (5). This element is positioned in front of Segment 1, near the visitor's path, and is illustrated in Fig. 5.13. It is a compact rectangular block measuring 0.86 × 0.40 × 0.40 m and weighing just

Fig. 5.13 a Luis observes block No. (5) located on the visitor's path. The arrow indicates the vertical slot; **b** the multiple cylindrical holes drilled in the slot

over 300 kg. The faces of the blocks are perfectly flat, with one of them featuring a vertical groove. This groove is 4 mm wide and 4 mm deep, and the vertical and horizontal inner faces meet at right angles.

The vertical slot contains approximately twenty cylindrical holes, each measuring 4 mm in diameter and 9 mm deep, with a spacing of 3.5 cm. It remains a mystery how the craftsmen of AD 600 were able to drill these holes. The cylindrical holes are all uniform, without any tapering, and their bottoms are flat, forming a right angle with the cylindrical walls. The tool or instrument utilized to create these holes remains unknown. Undoubtedly, these enigmatic features contribute to the rumors of extraterrestrial civilization unless the block itself is a result of a geopolymer or similar technique, which is what we precisely aim to demonstrate.

Moving on to the renowned "H" blocks, labeled as No. 6, these blocks are the true stars of Pumapunku. I previously mentioned them in Chap. 3 (Fig. 3.4). Figure 5.14 showcases the magnificence and perfection of these structures. Besides the photograph (Fig. 5.14a), we can observe the meticulous drawings by Stübel and Uhle (Fig. 5.14b), published in 1892, which highlight the geometric precision of these blocks down to the millimeter.

These sculptures stand at approximately 1 m in height, 1 m in width, and 0.5 m in thickness. They are estimated to weigh around 700 kg, and their vertical and horizontal surfaces are all precisely angled at sharp right angles. There are no visible rounded edges that would indicate the use of tools. Therefore, we endeavor to comprehend the techniques and methods employed in their creation.

Let us now turn our attention to the rectangular table, labeled as No. 7. This slab, almost square in shape, is positioned near the western corner of Segment 2 (Fig. 5.15). It measures 3.64 m in length, 3.30 m in width, and has a thickness of 0.45 m, weighing approximately 15 tons.

According to Table 5.1, its external dimensions are similar to those of the Puerta del Sol. The surface of the slab is flawlessly flat and smooth and possesses a hard texture on top, reminiscent of a finely ground and polished granite marble. It almost

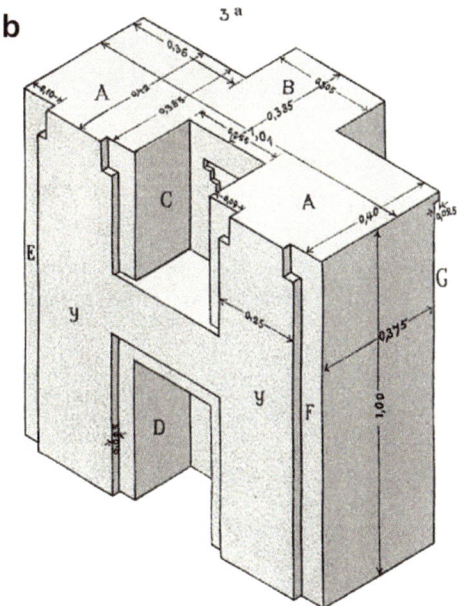

Fig. 5.14 **a** H-sculpture, No. (6); **b** the drawing by Stübel and Uhle (1892)

Fig. 5.15 Slab No. 7 near the western corner of Segment 2

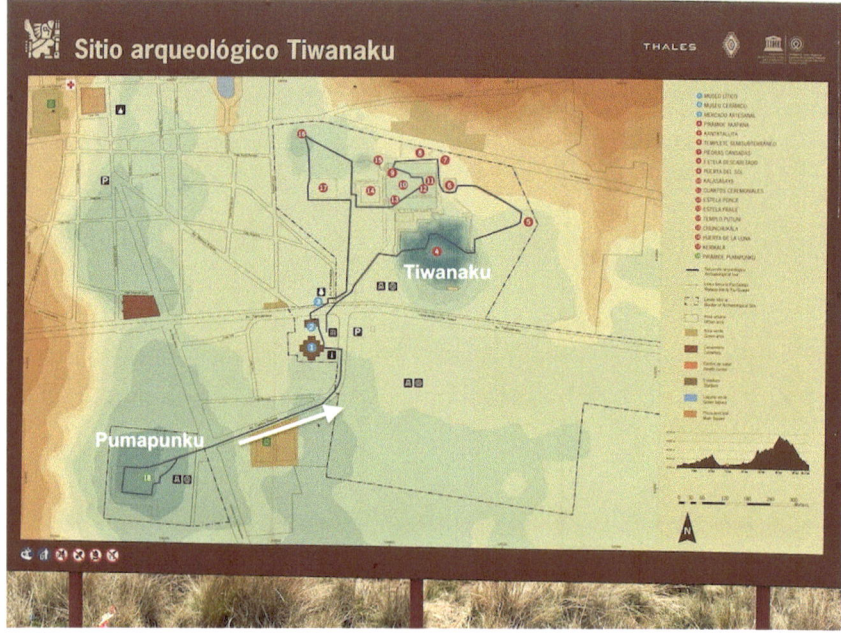

Fig. 5.16 Photo of the informative board of the archaeological site of Tiwanaku comprising Pumapunku on the left and Tiwanaku on the right

resembles a modern colossal piece of machinery. The details of this slab remain largely unknown, but its underside reveals several rectangular grooves, each 18 cm deep, clearly visible in Fig. 5.15.

This observation concludes our initial visit to Pumapunku. I am eagerly anticipating further exploration and delving into specific aspects of this fascinating site. The team will return in three days, to continue our study and gather samples whenever possible. Now, let us shift our focus to Tiwanaku, located on the other side of the road (Fig. 5.16).

5.3 Tiwanaku

The Tiwanaku site encompasses a much larger area compared to Pumapunku. While the Akapana pyramid is taller, it does not particularly stand out and falls within the average height of other pre-Columbian monuments. In that regard, Pumapunku with it small pyramid is an exception.

Our visit to the site will not follow the suggested route shown on the panel in Fig. 5.16, with a total of 14 designated points to explore, numbered 4–17 in Fig. 5.17. These include the Akapana Pyramid (4), Kantatallita (5), Semisubterranean Temple (6), Tired Stones (7), Stela Descabezado (8), Sun Gate (9), Kalasasaya (10), Ceremonial Quarters (11), Stela Ponce (12), Stela Fraile (13), Putuni Temple (14), Chunchukala (15), Moon Gate (16), and Kerikala (17).

Fig. 5.17 Our expanded photo of the informative board with added specific details: (4) Akapana and arrows indicating the foundations made of large stones; (9) Puerta del Sol; (10) Kalasasaya enclosure; (16) Puerta de la Luna

Fig. 5.18   General view from the Puerta de la Luna with the Puerta del Sol (Sun Gate)

Before we embark on our visit, let us take a moment to gain a general overview of the site. To do so, our team will go directly to the Moon Gate—Puerta de la Luna (16) and approach the circuit in reverse. It may not be the most conventional approach, but as we have seen in Pumapunku, this perspective allows us to understand the context and the surrounding environment better. Figure 5.18 provides a photographic representation of the main points of interest.

5.3.1   Puerta de la Luna

The Puerta de la Luna, also referred to as the Moon Gate, is another remarkable structure, though smaller in size (Fig. 5.19). Various authors widely agree that the gate is not in its original location. According to older texts, it was originally situated next to the Puerta del Sol and was later moved to its current position on the mound. Some believe that its purpose was to mark the entrance to a children's cemetery. The dimensions of the Puerta de la Luna can be found in Table 5.1, alongside those of the other gates. It is also apparently made of volcanic rock, and this feature is of interest to us. It has a frieze in its rear part, which is intact, and it does not possess any of the small holes mentioned above in Sect. 5.2.2 and Fig. 5.6. This means that it has not been covered with gold. Its right-hand jamb has a foreign geological inclusion that we might be able to study during our next expedition.

5.3.2   The Kalasasaya Temple

It consists of large red sandstone pillars and a combination of different types of masonry, forming a rectangular ensemble measuring 120 m by 130 m. I presented it in detail in Chap. 4 (Figs. 4.1 and 4.2), and it is depicted here in Fig. 5.20. However, some specialists question the accuracy of the reconstruction of this wall, suggesting

Fig. 5.19 Puerta de la Luna; the arrow shows the inclusion featured on the right

that it does not align with archaeological evidence. While the pillars are authentic, the majority of the masonry was constructed using small stone blocks sourced from outside the site. This setting is clearly visible in Fig. 5.20c.

Additionally, there is even a water channel, resembling a gargoyle or gutter, located at the upper part of the wall. It is known that the site has been used as a quarry for several centuries. Only the pillars themselves are original, and they hold our interest. They are constructed from red "sandstone," similar to the materials used in Pumapunku and Akapana.

These red "sandstone" pillars have withstood the test of time, enduring 1400 years of erosion caused by the harsh climate of the Altiplano. As a result, sheets or plaques have detached from the surface of the rock, a phenomenon clearly visible on the pillar (b) in Fig. 5.20, as well as on the large pillar (c). We observe a similar occurrence here to what we witnessed at Pumapunku. See further at the end of this chapter.

Within the Kalasasaya temple, there are several notable elements, including two statues known as Ponce's Monolith (Fig. 5.20) and the Fraile Monolith. The former is crafted from what resembles the volcanic rock andesite, similar to the composition of the 'H' blocks in Pumapunku. At the same time, the latter is made from two-colored red and yellow soft sandstone-like material. I mentioned the Fraile statue in Chap. 3, Fig. 3.4.

Fig. 5.20 Kalasasaya enclosure; **a** corner near the Puerta del Sol; **b** Corner opposite to (**a**); **c** Heteroclite filling with gargoyle (or gutter), arrow; **d** Ponce statue

5.3.3 The Puerta del Sol (Sun Gate)

Tourists from all corners of the globe flock to this site to marvel at the renowned Puerta del Sol, located in the remote part of Kalasasaya, in the northwest corner of Fig. 5.18. All authors widely acknowledge that the gate is not in its original position. It was undoubtedly moved from Pumapunku and suffered damage during transportation. It is worth noting that Cieza de León does not mention it in his text (Chap. 4, Part 1), indicating that it was likely not present in Tiwanaku in AD 1549, perhaps in Pumapunku.

Fig. 5.21 Puerta del Sol, a monolithic gate made of material fabricated with volcanic rock andesite. **a** Front and **b** Rear parts

In his book, *"The Stones of Tiahuanaco: Study of Architecture and Construction,"* architect-historian Protzen and Nair (2013, p 149) remark, "(…) *Whether the present location of the gateway is its original one is open to debate. It stands there eccentrically placed and incongruously isolated, with no obvious relationship to its immediate surroundings. Historical documentation claims that Marshall José Antonio Sucre, who was impressed by the ruins of Tiahuanaco and later became president of Bolivia, ordered the local authorities on February 6, 1825, to re-erect the Gateway of the Sun. This order, however, does not specify the gateway's original location or the place it was resurrected, if indeed the order was followed (…)"*.

Therefore, it had been moved before 1825, and it likely sustained damage during that time. It was later restored in 1908 by Arthur Posnansky, who was overseeing the site at the time. The frieze of the gate remains remarkably intact, devoid of the characteristic small holes found in Pumapunku, which suffered from degradation. As we learned in Sect. 5.2.2, the absence of gold leaf, and thus the lack of small holes, has contributed to its preservation. The jambs of the Puerta del Sol are wider than those of Pumapunku, giving it a more imposing appearance. We can compare its dimensions with the other gates in Pumapunku and Tiwanaku in Table 5.1. In Fig. 5.21, we can observe that the front part of the Puerta del Sol is well-maintained compared to the rear part, which is covered with lichen in its upper half.

The Puerta del Sol, like the structures in Pumapunku, is constructed from material fabricated from andesite volcanic rock, as shown later in this book. Since the nineteenth century, the front part has been replicated numerous times, often using plaster or paper. I recall seeing a replica of the entire monument at the Musée de

Fig. 5.22 a The base with the pillars covering the height of 4 red sandstone blocks; **b** To the left of the opening, the arrow shows the structure illustrated in Fig. 5.23

l'Homme, Porte Chaillot, in Paris during the 1960s. As a result, it was regularly cleaned. However, the same cannot be said for the rear facade, which features recesses and chiseling reminiscent of the gate at the entrance to Pumapunku (Fig. 5.3).

5.3.4 Akapana Pyramid

The Akapana pyramid follows a precise alignment with the cardinal points, measuring 194 m in length from east to west and 182 m from north to south, with a height of 16 m. This stepped monument consists of seven platforms stacked on top of each other. Each level was compacted using a mixture of clay and sandstone rubble. Initially, I did not find it particularly intriguing and did not prioritize it for in-depth study like the other monuments in Tiwanaku. However, Ralph had taken numerous photographs, and we ended up with a collection of 529 pictures. Over two and a half years later, as I started to write this chapter, I revisited those images. I recalled the passage in Cieza de León's text about Tiwanaku, where he mentioned, *"(…) Near the main buildings is a handmade hill, built on large stone foundations (…)."*

The base of the Akapana pyramid showcases impressive blocks of stone, similar to the red "sandstone" seen in Pumapunku. We can observe them in Fig. 5.22, specifically in the restored area of the base indicated by arrows in the plan depicted in Fig. 5.17. Some of these pillars are massive, standing nearly 2 m tall, and they consolidate four rows of blocks that are precisely joined with sharp accuracy (Fig. 5.22a).

However, the real surprise emerges from Fig. 5.22b, located to the left of the opening in the base. Here, we encounter a structure featuring a distinctive rectangular "key" shape, assembled in a manner typical of pre-Inca constructions (Fig. 5.23).

Fig. 5.23 Base of the Akapana with the arrangement of red "sandstone" blocks with sharp joints and typical rectangular "key" dating from AD 600

This revelation caught me off-guard because I had not come across it during my bibliographical research. This specific area at the base of the pyramid had only recently been cleared and had not been the subject of any prior publications.

5.4   Summary of the First Visit to Pumapunku/Tiwanaku

The first visit to Pumapunku/Tiwanaku has been incredibly enlightening, providing us with valuable insights for our investigation into the geological or artificial nature of the materials used. Let us summarize these discoveries:

Fig. 5.24 Erosion patterns of the lamellar type: **a** Vertical pillars of the Kalasasaya, Tiwanaku; **b** Horizontal surface of the Pumapunku megalith, segment n. 2

(a) The vibrant red color of what is called "sandstone," as pointed out by geologist Luis, appears to be unnatural and raises questions about its origin.

(b) The erosion patterns on both the vertical pillars of the Kalasasaya and the horizontal surfaces of the Pumapunku monoliths exhibit a lamellar (sheet or plate) type, as depicted in Fig. 5.24. This erosion pattern suggests the presence of a red geopolymer made from sandstone created through sedimentation. It is essential to determine if this phenomenon is also found in natural sandstone blocks.

(c) The gates and carved blocks, seemingly made of andesite rock, display numerous small cylindrical holes, numbering in the dozens or even hundreds. It is highly improbable that these holes were created using conventional tools. This finding hints at the possibility of a "malleable" stone technique, potentially related to geopolymer methods.

These clues indicate that we are making progress in our research. Tomorrow, our team will visit the first of the three red "sandstone" geological sites, located approximately 11–12 km south in the Quimsachata Mountains.

Regarding the extraction site for the andesite material, which is the Cerro Khapia volcano, I have not found a comprehensive study pinpointing its exact geological source. My investigation will depend on synthesizing existing knowledge contributed by various authors. Nonetheless, I am optimistic that the outcome will be exceedingly positive.

References

Gara TA, Davidovits J, Davidovits F (2020) Considering certain lithic artifacts of Tiahuanaco (Tiwanaku) and Pumapunku (Bolivia) as geopolymer constructs. Geopolymer Archaeol 1: 44–53. https://doi.org/10.13140/RG.2.2.36569.75366/1

Gara TA (2016) Considering five artifacts at Tiwanaku and Puma Punku. Available on https://www.academia.org

Protzen JP, Nair S (2013) The stones of Thiahuanaco: study of architecture and construction. Monograph 75, Cotsen Institute of Archaeology Press, UCLA, Los Angeles

Stübel A, Uhle M (1892) Die Ruinenstätte von Tiahuanaco im Hochlande des alten Peru: eine kulturegeschichtliche Studie. Karl W. Hiersemann, Leipzig. https://doi.org/10.11588/diglit.21775#0004

Vranich A (2018) Reconstructing ancient architecture at Tiwanaku, Bolivia: the potential and promise of 3D printing. Heritage Sci 6:65

Chapter 6
Study of the Selected Red Sandstone Geological Sites

Ponce Sangines' geological report—Looking for other potential sites—Visit to geological site No. 1: Kaliri/Quebrada de Kausani—Visit to geological site No. 2: Cerro Amarillani—Visit to geological site No. 3: Kallamarka—Why is Kallamarka such a special place?—First conclusions on the geological study of the red sandstone.

In this chapter, we will examine the geological origin of the red material used in the megalithic platforms (1)–(4) of Pumapunku that resembles sandstone. I introduced my son Frédéric at the beginning of Chap. 2, who was busy transcribing the report by Carlos Ponce Sangines' team titled "Origin of the sandstones used in the pre-Columbian temple of Pumapunku (Tiwanaku)," which was published in 1971. I had approximately two months to thoroughly study and comprehend this report in order to present Luis and Ralph with the plan for their geological study.

6.1 Ponce Sangines' Geological Report

Ponce Sangines' study primarily focused on six geological sandstone deposits located in the Quimsashata massif, situated south of Tiahuanaco. These deposits are visually represented in Fig. 6.1. The "Mo" samples mentioned refer to the study conducted with Gerardo Mogrovejo Terrazas, published in 1970 (Ponce Sanginés and Mogrovejo 1970).

(A) Cerro Chulipa, samples A1–A3, located 10,000 m in a straight line from Tiwanaku.
(B) Quebrada Rio Quimsachata, samples B4–B6, Mo 470 and 526, at a distance of 9100 m.
(C) Quebrada Rio Jarpajawira, samples C7–C9, Mo 405 and 406, approximately 7500 m away.

J. Davidovits, *Ancient Geopolymers in South America and Easter Island*,
SpringerBriefs in Earth Sciences, https://doi.org/10.1007/978-3-031-75336-7_6

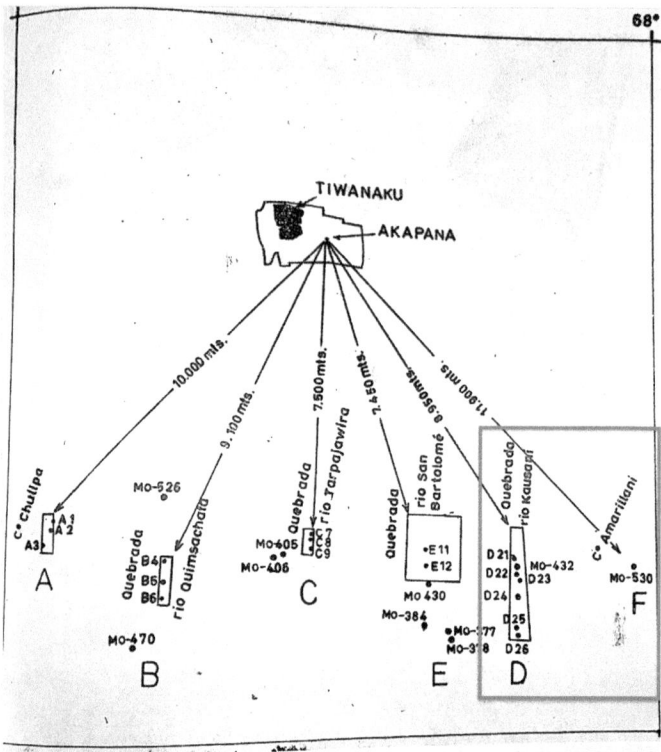

115. Distancias desde Akapana, volumen mayor de la urbe tiwanacota, a las quebradas
donde existe arenisca roja. Se ha marcado también la ubicación de las muestras ex-
traídas para la presente monografía.

Fig. 6.1 The six geological sites studied by Ponce Sangines' team in the report "Origin of the
sandstones used in the pre-Columbian temple of Pumapunku (Tiwanaku)," 1971

(D) Quebrada Rio Kausani, samples D21–D26, Mo 432, situated at a distance of
8950 m.

(E) Quebrada Rio San Bartolomé, samples E11 and E-12, Mo 430 and 384, located
around 7450 m away.

(F) Cerro Amarillani, sample Mo 530, positioned 11,900 m from the site.

These various locations provide crucial information regarding the potential
sources of the red sandstone utilized in Pumapunku. In Ponce Sangines' report,
the primary document is titled *"Comparative Petrographic Study of the Pumapunku
Sandstones,"* conducted by Arturo Castaños Echazú, a Ph.D. in geology and the Dean
of the Faculty of Geology at the Universidad Mayor de San Andrés (translated from
Spanish):

Pages 210–212 of the report state: "(…) *The petrographic comparison with the samples of the ravine designated with the letter A leads to ruling it out as the site of origin or quarry of the Pumapunku blocks (…) In addition, the ones from gorges B and C have to be ruled out for a number of reasons. Although those collected in ravine E offer some similar petrographic particularities, they cannot be fully identified with those from Pumapunku (…) On the other hand, the samples named D-21, D-22, D-23, D-24, D-25, and D-26 show, in my opinion, an undoubted petrographic similarity as far as mineralogy, modal analysis, morphoscopy and size are concerned. Consequently, the D ravine (quebrada), through which the Rio Kausani flows and which is about 8 km from Pumapunku in a direct line, is taken as the best one.*

(…) In conclusion (…) The comparison is in favor of those of group D from the Rio Kausani (fig 129) (Quebrada Kausani), given that there is evidence of petrographic overlap with those of the archaeological site. Probably, the stone material for the construction of the aforementioned Tiwanaku monument was brought from there in the past. We must now turn to the chemical analysis and the mineralogy of the clays, to which the following pages are devoted, and which were carried out by the co-authors Urquidi and Avila, to determine whether or not the obtained results are coherent (…)".

Following Arturo Castaños Echazú's advice, I thoroughly examined the mineralogical and chemical analyses. However, I must note that Urquidi and Avila, the authors mentioned, also reference group D, Quebrada Kausani. Nevertheless, I have a distinct feeling that their inclusion of this site is more to follow mainstream opinion rather than rooted in scientific rigor.

The Kausani site has long been recognized in the region due to its unique geological characteristics, which strongly support its significance. Locally, it is referred to as Kaliri or Kalari, as depicted in Fig. 6.2. This site showcases numerous rectangular sandstone blocks ready to be utilized as construction material. Figure 6.2 is a reproduction of Fig. 128 from Ponce Sangines' 1971 report. The caption beneath it in the report describes the process of mechanical weathering disintegration and the separation of blocks through diaclasis.

The site at Kaliri/Kausani presented an ideal location for sandstone exploitation due to the presence of these naturally rectangular blocks. However, there is no evidence to suggest that ancient stonemasons actually frequented this site. The primary drawback of Kaliri/Kausani is its high altitude, exceeding 4200 m, making it relatively challenging to access. Our team has experienced this firsthand, and topic 6.3 provides a glimpse of the difficulties encountered. Despite this, Bolivian archaeologists, accustomed to the remarkable achievements of their ancestors, were undeterred.

Furthermore, American anthropologists like Janusek et al. (2013) consider the Kaliri/Kausani site to be a quarry. Janusek writes on page 79 about the results obtained by Ponce Sangines' team: "*The authors did not consider this field (Kausani) of sandstone blocks the result of human activity. Our research determined that it was an anthropogenic production.*" He further elaborates on page 84, "*We located only one positive source of sandstone for Tiwanaku in the Kimsachata range; that is, only*

Fig. 6.2 Reproduction of the Fig. 128 displayed in the Ponce Sangines' report mentioned in Fig. 6.1, with the blocks resulting from natural geological diaclasis

one source where partially carved stones had been left behind. Kaliri/Kausani was likely the major source of Tiwanaku's sandstone" (Janusek et al. 2013, p. 79).

In other words, Janusek and his colleagues based their selection solely on the presence of more or less regular blocks, disregarding deposits that lacked this geological characteristic. This is a logical choice for those who believe the blocks were hewn, but it is not my perspective. As I mentioned earlier, I studied the results obtained by the new colleagues of Ponce Sangines, who took over from geologist G. Mogrojevo (the Mo samples in the list). I compared the chemical analyses conducted on the twenty samples from the megalithic terraces (1)–(4) with the six samples from the Kausani/ Kaliri geological site (D). I discovered clear differences that raised significant doubts in my mind.

Now, let us examine the results from the report titled *"Geochemistry of the Pumapunku sandstone"* by Fernando Urquidi Barrau, a geological engineer. On page 231, he states, *"Only three elements were analyzed in the group of major chemical elements, specifically silicon, iron, and calcium. The analyzed iron and calcium are believed to belong to the cement that binds the major minerals. The minor elements of the rocks were analyzed using spectrochemical methods, which determined the trace elements semi-quantitatively, with results reported in parts per million (ppm).*"

The analysis of lime (CaO, calcium) and iron content, as highlighted by Urquidi, reveals a significant disparity in the quantities of binder (cement) present. The Pumapunku sandstone contains approximately five times more lime CaO (mean value 1.70%) than the sandstone found at Kausani/Kaliri (mean value 0.36%). Additionally, notable differences arise in the analysis of trace elements, specifically for three chemical components: boron (B), barium (Ba), and copper (Cu). The samples they analyzed from Pumapunku sandstone resulted in 100 ppm boron (B), 230 ppm barium (Ba), and 20.5 ppm copper (Cu). These values are much higher than those found in the natural sandstone of Kausani/Kaliri, namely below the detection limit for boron (B), 6 ppm barium (Ba), and 10 ppm copper (Cu).

Based on these findings, we can confidently conclude that the red sandstone of Kausani/Kaliri does not match the red "sandstone" observed in the Pumapunku terraces. Consequently, the conclusions drawn by Ponce Sangines' new team, along with those of the American anthropologists and archaeologists, are erroneous. This perspective leads us to question the whereabouts of the geological source responsible for the Pumapunku sandstone.

6.2 Looking for Other Potential Sites

Upon reviewing the data, I noticed a significant oversight by the new team of Ponce Sangines and Janusek. They failed to consider the previous analyses conducted on the Mo-430 sample from the Cerro Amarillani site (F), which is located further to the east of (D) in Fig. 6.1. Intrigued by this discovery, I decided to gather more information on this specific matter. Searching online, I obtained a geographical map of the region and discovered that Cerro Amarillani is in close proximity to a village called Chuñuni.

As I visualize the extensive stone carving and quarrying activities, it becomes apparent that they must have involved the labor of hundreds, if not thousands, of families over an extended historical period. It is essential to remember that, apart from Pumapunku, the Tiwanaku site itself contains a substantial quantity of red sandstone blocks and pillars, as described in the previous chapter. Such large-scale technological endeavors would undoubtedly leave behind archaeological traces and remnants.

However, when I refer to these traces, I am not simply alluding to the few scattered sandstone blocks in the valley that American anthropologists often cite as evidence to support their claims. This topic will be further explored in the next chapter, where we discuss the abandoned andesitic rock blocks on the paths of Cerro Khapia volcano, situated on the other side of Lake Titicaca.

When I mention archaeological traces and remnants, I am explicitly referring to the existence of villages and populated areas where the families of the numerous workers who exploited the red sandstone deposit once resided. Considering the duration of such activities and the construction of monuments, temples, and cemeteries, it is reasonable to expect that these villages still exist. However, the reality is that, thus far, no archaeologist or anthropologist has sought to establish a connection between the geological sites and the presence of archaeological remains. To the best of my knowledge, there is no ancient village associated with Kausani/Kaliri, which, in a way, aligns with my previous conclusion that this site is not the correct source from which the Pumapunku red sandstone was quarried and worked.

The name Chuñuni, associated with Cerro Amarillani, struck a chord with me as I recalled coming across it in Janusek's book, "*Ancient Tiwanaku*" (Janusek 2008). In fact, I found references to it in two passages. The first was on page 73, Fig. 3.4, where it appeared on a map titled "*Chiripa and other major Formative sites in the Southern Lake Titicaca Basin.*" The second mention was on page 91, where Janusek stated that during the Late Formative period (around AD 400), Tiwanaku served as the hub for a society that encompassed a significant portion of the Tiwanaku Valley. Other nearby sites, including Kallamarka and Chuñuni, were mentioned, located several kilometers to the east.

Drawing from the map and using resources like Google Earth, I discovered that these two villages were adjacent to each other, located at the foot of the hills of the Quimshasata massif (Fig. 6.3). On the map, one can even discern the remains of an old road or trail connecting these villages directly to Tiwanaku.

There was no doubt in my mind that these were the places to explore further: Cerro Amarillani at Chuñuni and a yet-to-be-determined site at Kallamarka.

To summarize, our three geological study sites are as follows, as depicted in Fig. 6.3:

1. Kaliri/Quebrada de Kausani
2. Cerro Amarillani/Chuñuni
3. Kallamarka/Kalla Marka.

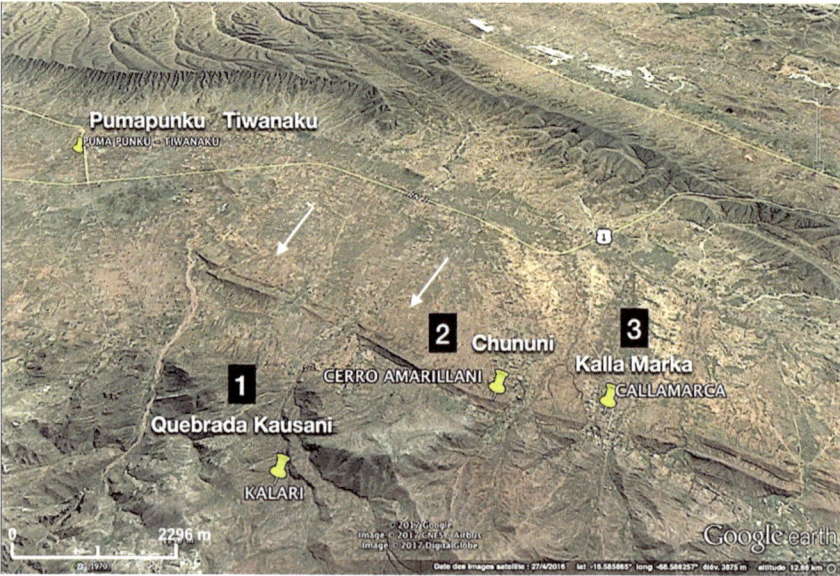

Fig. 6.3 Location of the three geological sites on Google Earth; the arrows point to the ancient trail connecting Tiwanaku and Chuñuni

6.3   Visit to Geological Site No. 1: Kaliri/Quebrada de Kausani

Although it may not be the ideal geological site, we must begin our exploration there in order to validate my reasoning. In Fig. 6.4, you can see the entrance to Quebrada del Rio Kausani as it appeared in 1970, alongside a photo taken by Ralph at the same location in 2017. To reach the Kaliri site, Luis and Ralph cross the new bridge and follow the path that winds along the steep slopes of Rio Kausani. They leave the driver with the hired 4 × 4 car under the tree in front of the bridge, and their ascent begins.

Fig. 6.4   a Quebrada del Rio Kausani in 1970 (Ponce Sangines); **b** photo of RD in 2017

Fig. 6.5 **a** The ascent to the Kaliri/Kalari site and the GPS track from 3910 to 4215 m, **b** and **c** Following the rocky path dotted with rocks and pebbles

Fig. 6.5 (continued)

Equipped with a GPS, Ralph traces their route, as shown in Fig. 6.5. Even for llamas, this steep track, bordered by a ravine and scattered with rocks and pebbles, proves treacherous. They must climb a 300-m elevation difference over a distance of 2.3 km, resulting in a slope of 13–15%. Ralph describes the climb as a physically demanding experience, mainly due to the high altitude of over 4000 m. Both Ralph and Luis find themselves catching their breath and taking frequent breaks. Despite Luis being accustomed to living at an elevation of over 2000 m in Arequipa, the altitude still affects him. Additionally, they must exercise caution to avoid injury, as a sprained ankle would hinder their progress.

Once they reach the summit (as shown in Fig. 6.6), Ralph and Luis imagine themselves as part of the team of stonemasons from the years 500–600 AD. They envision the daunting task of loading blocks weighing between 2 and 5 tons onto sleds, which llamas would then drag down. This operation would have been incredibly perilous for both humans and animals, particularly during the descent.

Upon their exploration of Kaliri, they come to a realization–no massive red sandstone megaliths weighing 100 tons or more, like those found in Pumapunku. This observation aligns with JP Protzen's findings (2013, p 176), who visited the site twice and stated that the precise location from which the colossal sandstone slabs in Pumapunku were extracted remains unknown. Other potential quarry sites they investigated at the foot of the Quimsachata range did not possess the necessary characteristics. Even if they had discovered such slabs, the question remains as to how they would have managed to transport them down the 15% slope and the trail depicted in Fig. 6.6.

Fig. 6.6 **a** Luis arrives at the top with the scattered sandstone blocks at Kaliri; **b** The arrow indicates the shepherd's house and its enclosure, built entirely with local stone

Based on their visit to the Kausani/Kaliri site chosen by Ponce Sangines' team in 1971, their first conclusion is clear–this is not the geological source of the red sandstone found in Pumapunku that they are seeking. To settle the debate once and for all, Luis selects a sandstone boulder and captures a rock sample that we plan to analyze in France. Notably, the color of the rock appears lighter than that of the sandstone found in Pumapunku (as depicted in Fig. 6.7).

Fig. 6.7 **a** The the sandstone boulder where the sample **b** was taken

6.4 Visit to Geological Site No. 2: Cerro Amarillani

Cerro Amarillani, located at the base of the Quimsashata range as depicted in Fig. 6.3, proves to be an interesting site. It is a surface fold, resembling a bar, and shares the same geological composition as Kaliri. Standing at around 30 m in height, it sits close to the earthy road, making it much more accessible than Kaliri. The GPS tracking in Fig. 6.8 reveals a short distance to reach the site.

Upon arrival, Luis notices that the sandstone here exhibits a vibrant red color. Additionally, there is an abundance of red clay that could have been utilized in pottery production. Another advantage of this location is the convenient transportation route for the stone materials. They could have been easily transported in a straight line

Fig. 6.8 **a** The ascent route on the slopes of Cerro Amarillani; **b** on the top, the sandstone boulder from where the sample was taken; **c** Luis examining the red clay layer

Fig. 6.8 (continued)

from the village of Chuñuni to Tiwanaku, traversing flat terrain. However, despite these favorable conditions, they do not find megalithic rocks of similar dimensions as those found in the Pumapunku terraces.

6.5 Visit to Geological Site No. 3: Kallamarka

On the third day, Luis and Ralph drive directly from La Paz to Kallamarka with the same 4 × 4 car and the same driver. They had an intriguing experience during their visit. They were surprised by the distinctive entrance to the village, marked by a welcome arch (see Fig. 6.9) with an inscription that translates to *"Welcome, Original Marka Callamarka."* As they explored the village, they observed its unique character and suspected it held a rich historical background. They sent me the photos, and I started a search on the Internet to understand what had been going on there.

Continuing their journey, they returned to the earth road that led them further up the mountain slopes. The route is marked by GPS in Fig. 6.10.

Luis carefully scanned the surroundings until he spotted a location that appeared to match the geological features they were seeking. They found rectangular sandstone blocks of a vibrant red color resting on layers of weathered and eroded sandstone (Fig. 6.11). Luis noted that besides the altered blocks, the most fascinating aspect was the layer of disintegrated sandstone, which seemed suitable for the geopolymeric reaction they were investigating.

Fig. 6.9 Kallamarka: **a** the entrance arch; **b** the brick-lined main street

In the video, Luis, the geologist, conducts an experiment to demonstrate the properties of the rock. By using a simple tool, he breaks the weathered rock easily, indicating its potential as a suitable material for geopolymers. The video emphasizes the contrast between the easy crumbliness of the weathered rock and the hardness of the unaltered rock (Davidovits 2019).

Fig. 6.10 **a** The GPS traces the route through Kallamarka and shows the selected geological site; **b** Continuing on the earth road, the arrow points to the site of the red sandstone deposit chosen by Luis, near the 4 × 4 car

Fig. 6.11 a The layer of
sandstone disintegrated by
erosion; **b** Luis easily breaks
the sandstone and takes
pieces for analysis

6.6   Why Is Kallamarka Such a Special Place?

I discovered fascinating information about the distinctive welcoming arch in Kallamarka (Callamarca) and its significance. It appears that the village holds historical
importance as a stop along the extensive Inca Trail, specifically on the Qhapaq Ñan
route. Through an internet search, I found an article reporting the visit of Irina Bokova,
the Director of UNESCO, to Callamarca on June 16, 2014 (Ultima-Hora 2014).

The article mentions that Bokova expressed her expectation that the Qhapaq Ñan, the network of pre-Hispanic Andean roads, would be inscribed as a World Heritage site within a week. She made this statement during her visit to the indigenous Aymara community of Callamarca (Kallamarka), where one of the sections of Qhapaq Ñan passes through in Bolivia. Bokova emphasized that once inscribed, the Qhapaq Ñan would belong to the entire world, not just the local community.

The Qhapaq Ñan, also known as the Main Andean Road, spanned an impressive 23,000 km and played a crucial role in the political and economic power of the Inca empire. It connected various production, administrative, and ceremonial centers. In the case of Callamarca, the village served as an intersection of roads along the Qhapaq Ñan route, as depicted in the map shown in Fig. 6.12.

The newfound understanding of the village's historical significance adds another layer of context to our exploration and research of Pumapunku. It highlights the interconnectedness of the region and its role within the broader Inca civilization.

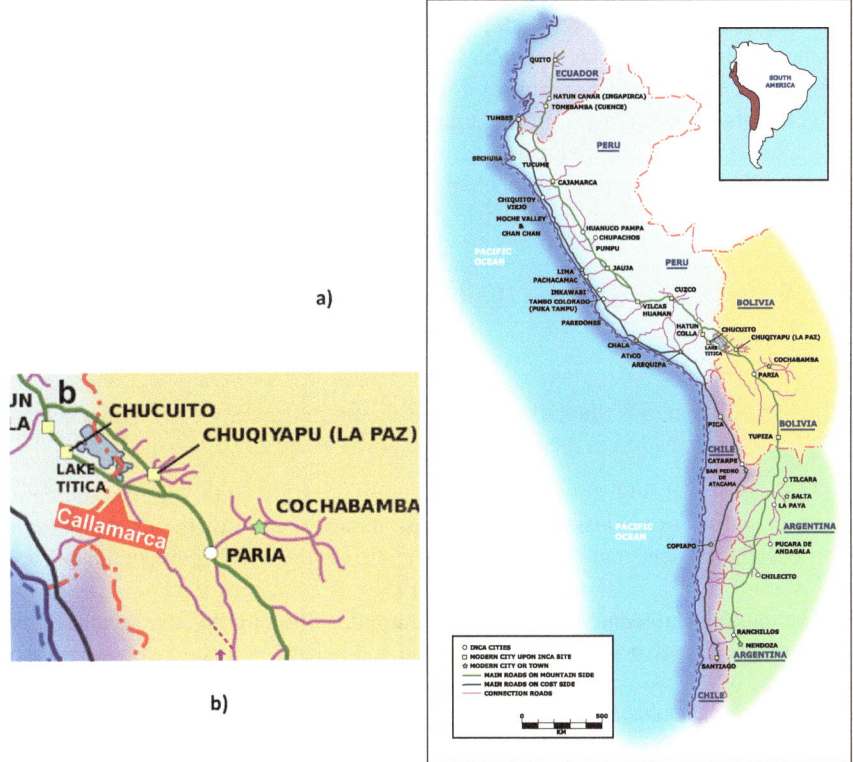

Fig. 6.12 a Qhapaq Ñan routes in Ecuador, Peru, Bolivia, Chile, and Argentina, in red Callamarca; (Adapted from Inca Road System, 2024); **b** The Callamarca junction on the Qhapaq Ñan route (green) and the secondary routes (purple)

The Spanish writing of the village, Callamarca, aligns with the name Kallamarka. It is interesting to learn that Callamarca was situated on the western part of the western branch of the Qhapaq Ñan, near Lake Titicaca. Moreover, it served as the starting point for two secondary roads: one connecting to the eastern branch of the Qhapaq Ñan and the other leading south towards the desert regions of the Salar de Uyuni and the Laguna Cachi, as detailed in Chap. 8.

Considering its historical significance, Callamarca/Kallamarka must have played a vital role due to its strategic location in the extraction of geological materials during the development and construction of Tiwanaku and Pumapunku centuries ago. The town's prominence and influence were likely a consequence of these factors.

However, it is unfortunate that during our team's visit in November 2017, there were no tourist signs indicating that Callamarca was a UNESCO World Heritage Site. This absence extended not only to the road from La Paz to Tiwanaku and Pumapunku but also to the entrance to the village itself. It appears that the welcoming archway and the main brick street were specifically prepared for the visit of the UNESCO Director-General three years prior. If these signs had been present, it would have immediately confirmed that Kallamarka was indeed one of the geological reference sites we were seeking in our study. The lack of clear indications regarding Callamarca's historical significance is a missed opportunity for visitors to recognize its importance and appreciate its connection to Tiwanaku and Pumapunku. Nonetheless, our research and exploration have allowed us to uncover the village's significance and its role in the larger context of the region's geological materials and historical heritage.

6.7   First Conclusions on the Geological Study of the Red Sandstone

Following our visit to Kallamarka and the observations made, we have gained some initial insights that address the questions we posed at the end of Chap. 5:

1. The site of Kaliri/Quebrada del Río Kausani does not align with the source of the megaliths used in the construction of Pumapunku. Through our exploration, it became evident that the extraction site differs from Kaliri/Quebrada del Río Kausani.
2. The natural sandstone color at Kaliri/Kausani is lighter compared to that of the sandstone found at Cerro Amarillani and Kallamarka, which closely resembles the hue of the sandstone-like material found in Pumapunku.
3. Based on our observations, it is interesting to note that the erosion patterns in the natural sandstone blocks at Kallamarka differ from the laminar erosion we observed in Pumapunku. This distinction raises the possibility that the red "sandstone" used in Pumapunku might be artificial.
4. Significantly, the site of Kallamarka exhibits a strong historical connection to Tiwanaku and Pumapunku. This finding serves as a valuable initial verification while we eagerly wait for the results of our scientific analysis.

With these preliminary findings in mind, our next objective is to investigate the origin of the distinctive rocky substance that resembles andesite used in the construction of Pumapunku's gates and H-shaped sculptures. We are determined to delve deeper into this matter and uncover the source of this remarkable material.

References

Davidovits (2019) Tiwanaku/Pumapunku megaliths are artificial geopolymers. Geopolymer Institute channel on YouTube. Available at: https://youtu.be/rf9qK9QTlq0?si=QdTDHV2MycYSSGUr&t=709. Accessed on 07 July 2024

Inca Road System. Available at: https://en.wikipedia.org/wiki/Inca_road_system. Accessed on 10 Aug 2024

Ponce Sanginés C, Mogrovejo TG (1970) Sobre el origen del material lítico de los monumentos de Tiwanaku. Publicación no. 21. Academia Nacional de Ciencias de Bolivia, La Paz

Ultima-Hora (2014) Available at: https://www.ultimahora.com/la-titular-la-unesco-espera-que-designen-patrimonio-mundial-qhapaq-nan-esta-semana-n804191.html. Accessed on 10 Aug 2024

Chapter 7
Geological Study of the Sites Selected for the Andesite Rock

There has been no published study by Ponce Sangines?—The *"piedras cansadas,"* the tired stones —The andesite volcanic sand found in Iwawe—*Carbunculus* and the gas pipes: the volcanic sand in ancient Roman mortar—Who transported the *piedras cansadas* and when?

Chapter 1 started with an exploration of the artificial nature of the alleged andesite sample obtained from Pumapunku. The revelation of its artificial composition took me by surprise, and its impact on the public was significant following the publication of our scientific article in the journal *Ceramics International* in January 2019 (Davidovits et al. 2019a, b).

The focus of our scientific expedition, which I had organized, and the work program I had devised for Luis and Ralph, was solely centered around investigating the megalithic blocks considered to be made of red sandstone. Although the study conducted by Ponce Sangines' team in 1971 served as an excellent starting point, I realized that I should have also prepared a geological agenda for the andesite rock. However, at that time, I had not come across enough material to develop such a list. The present chapter serves as a continuation of our research, building upon a recently published scientific article (Davidovits and Davidovits 2020) in the *Journal of Geopolymer Science Applied to Archaeology.*

7.1 There Has Been no Published Study by Ponce Sangines?

I was aware that the prevailing belief pointed to the andesite's geological origin in the volcanic region of Cerro Khapia, Peru, situated across Laguna Huiñaymarca, which is connected to Lake Titicaca (Fig. 7.1). In the nineteenth century, several archaeologists made notable discoveries at this site, identifying potential sources for the stones (interpreted to be volcanic) used in the monuments of Tiwanaku and Pumapunku.

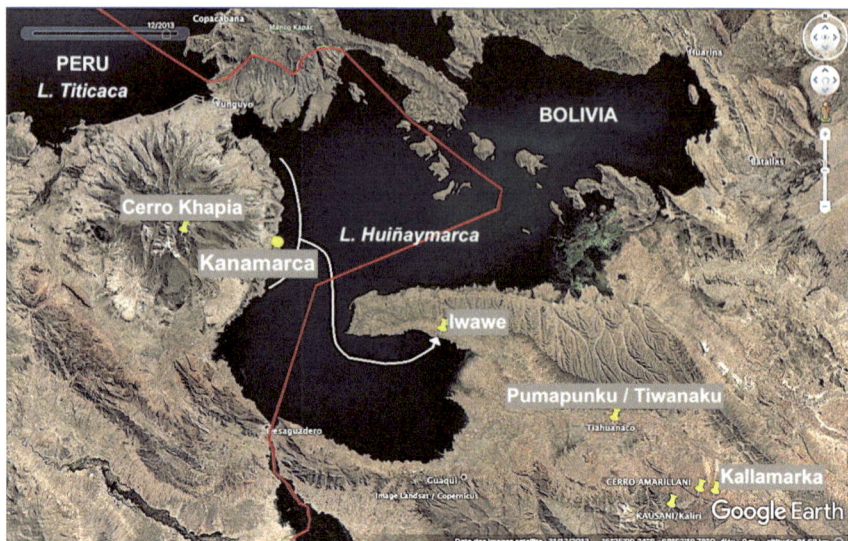

Fig. 7.1 Map showing the border between Peru and Bolivia, the position of Pumapunku/Tiwanaku in relation to the Cerro Khapia volcano, the small lake Huiñaymarca at the end of Lake Titicaca

For instance, Stübel and Uhle published their comparative analyses in 1892, drawing a strong correlation between the stone used in the Puerta del Sol and the geological rock found at the base of the Cerro Khapia volcano. This finding provided compelling evidence of a connection between the monument and its geological source.

However, I must acknowledge that I did not have a specific itinerary to propose to Luis and Ralph. The exploratory work conducted by Ponce Sangines in 1968, focusing on the geological origin of the andesite rocks, was incomplete. Ponce Sangines himself explained this limitation in his report: "*It, therefore, seems critical at this point to focus on the possibility of Kapira (Khapia) quarry. It has not been possible for us to mention it nor to make any field reconnaissance, as it is located near the border between Peru and Bolivia, on the territory of our brother country, where, for obvious reasons, there are too many difficulties for even a cursory exploration, and we hope that they will be overcome very soon for the benefit of science.*" (Mille and Sangines 1968, p. 37).

The map in Fig. 7.1 clearly indicates the border between Bolivia and Peru, highlighting the geographical context and illustrating the challenges faced in conducting further explorations due to the border location.

7.2   The "*Piedras Cansadas*", The Tired Stones

I am facing a dilemma. Despite our scientific discovery suggesting the artificial nature of the "andesite" rock at Tiwanaku/Pumapunku, there is a perplexing contradiction that cannot be ignored. The presence of rectangular volcanic blocks, known as the "*piedras cansadas*" or tired stones, scattered along the shores of both sides of the lake raises questions.

These blocks, weighing between 5 and 10 tons, have been described by countless travelers and archaeologists since the nineteenth century. They have attempted to unravel the mystery of how these massive volcanic rocks were transported from the slopes of Cerro Khapia volcano to the lake shores. The prevailing theory suggests that they were loaded onto rafts made of totora (tule) reeds, ferried across the lake, unloaded on the opposite bank, and then transported overland to Tiwanaku/Pumapunku. Figure 7.2a captures some of these andesite blocks on the Peruvian shores near Kanamarca, while Fig. 7.2b showcases others located on the Bolivian shores near Iwawe.

However, I find it puzzling to comprehend why the builders of Pumapunku would go through the arduous task of transporting 10-ton blocks just to cut them into smaller pieces for their famous "H" sculptures. Logically, it would have been much more straightforward to move smaller blocks weighing between 300 and 700 kg using small sleds and rafts instead of undertaking such a monumental endeavor. This lack of coherence leads me to believe that there must be another explanation, one that I am determined to uncover.

After the discovery of the andesite "*piedras cansadas*" at Iwawe on the Bolivian side of the lake, scholars such as Mille and Sangines (1968) and his Bolivian and American archaeological colleagues Isbell and Burkholder (2002), Janusek, Vranich, Isbell (2013), and Protzen proposed a scenario involving the crossing of the lake by boat from Kanamarca. Iwawe served as the port where these blocks were unloaded from the rafts and prepared for transport to Tiwanaku/Pumapunku. Refer to the map in Fig. 7.1 for a visual representation. However, for reasons unknown, these blocks

Fig. 7.2   "*piedras cansadas*", andesite boulders; **a** Peruvian shore at Kanamarca; **b** Bolivian shore at Iwawe. *Credit* Bardales Vassi

were not transported further to Tiwanaku/Pumapunku, leaving an intriguing puzzle waiting to be solved.

7.3   The Volcanic Andesitic Sand Found in Iwawe

This particular scenario gained support from William Isbell, a prominent North American anthropologist specializing in Andean ceramics, who conducted excavations at Iwawe between 2000 and 2002. Iwawe is a modest mound, standing no more than 3 m tall, and consists of nine archaeological layers labeled (I) to (IX), as depicted in Fig. 7.3.

In the first report published by Isbell and Burkholder (2002, p. 212), one reads: *(…) Stratum V is 15 cm deep, and it is frequently broken by a big pit excavated through it. (…) Stratum V demand interrogation. It is virtually sterile, and its distinctive material does not seem to be midden accumulated through domestic activities. At the suggestion of James B. Richardson, we submitted samples of Stratum V soil to volcanologist Richard Naslund (personal communication January 1994) of Binghamton University Department of Geology whose thin section microscopic examination conclusively revealed volcanic pumice. Unfortunately, though, the only soil sample available to us in the USA came from a flotation heavy fraction. Consequently, the relative sizes of particles in the original matrix could not be determined. Shanaka de Silva (de Silva and Francis 1991: 138–155 personal communication), who will collaborate with the Iwawe research team in the future, informs us that Stratum V could be volcanic ash rained down on the Altiplano following an eruption in the prehistoric past. In that*

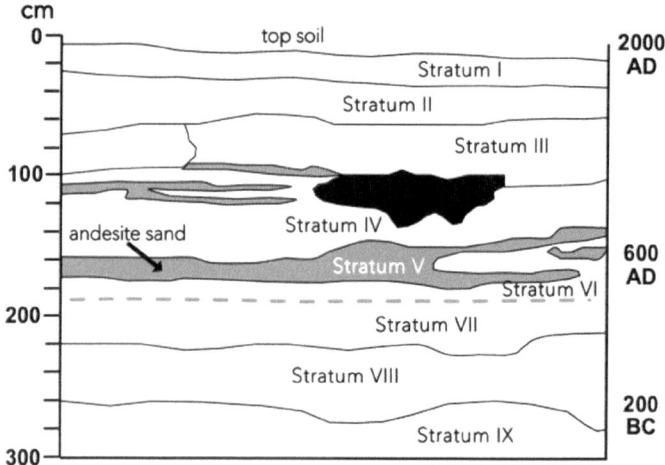

Fig. 7.3 Stratigraphy of the Iwawe mound revealed in the archaeological excavations, adapted from Isbell and Burkholder (2002). Stratum V (andesite sand) is represented in grey, and charcoal in black

case, the most likely source is the volcano known as Cerro Quemado, 250 km to the southwest of Iwawe. (...) However, volcanic ash is very durable and it can be blown by the wind for long distance. So the Iwawe ash may originate in some other primary tephra.

While this analysis provides some insights, it can be confusing and not entirely accurate due to the limited nature of the volcanic ash sample available to the American geologist. Despite this limitation, it is concluded that Stratum V is composed of ash of volcanic origin rather than a pile of domestic waste. The question remains: how did this volcanic ash find its way to Iwawe?

Following his new study on the site, Isbell further discussed the subject in 2013 (Isbell 2013, p. 175). He noted that Iwawe gained initial recognition as Tiwanaku's port, where immense andesite stone blocks were unloaded before being transported to the city. Additionally, Iwawe is an ancient mound composed of approximately 2.5 m of residential debris located along the shores of Lake Titicaca. Excavations at the site revealed nine distinct layers, with the lowest layer being sterile soil that had been shaped into raised fields for cultivation prior to human occupation. Stratum V, in particular, played a significant role in dividing the sequence. This layer consisted of andesite sand, likely a byproduct of extensive work involved in shaping the imported stone blocks. Isbell (2013) suggests that this gritty layer represents a period of intense construction using andesite, indicating a major construction phase at the capital.

Therefore, upon closer examination, the archaeological excavations revealed that Stratum V consisted of volcanic andesitic sand, not volcanic ash. In terms of granulometry, volcanic sand (referred to as "lapilli") is coarser than ash, which is comparable to mud-sized sediments. Isbell (2013) concluded that this layer was a result of the meticulous sculpting of the blocks that had arrived from the Cerro Khapia volcano. Yet, it is essential to consider that stone carving typically produces shards, varying in size from large to small pieces, rather than sand. To support this observation, I studied the internet presentation of a book published in 2018 by the Museo Nacional de Etnografía y Folklore (MUSEF—in English, National Museum of Ethnography and Folklore) of Bolivia. The book, titled *"Almas de la Piedra,"* features a photo on page 54 illustrating the andesite exploitation in Viacha, a town situated between Tiwanaku and La Paz. This photo, shown in Fig. 7.4, clearly demonstrates that the process of working with andesite blocks generates a substantial amount of waste material in the form of large and small stone pieces but almost no sand. This raises further questions about the origins and composition of the andesite sand found in Stratum V at Iwawe.

Isbell (2013) indeed does not mention the presence of andesite splinters in Stratum V, as he identifies it as being comprised of sand. However, he does possess the ability to distinguish between andesite sand and crushed andesite rock. This distinction is evident in Fig. 7.3 of the stratigraphy, which displays a thin layer of crushed rock labeled as a *"discontinuous stratum of crushed rock"* between Stratum VI and Stratum VII.

Furthermore, Isbell provides a more detailed description of the Iwawe excavation. In the deeper layers, precisely strata VI through VIII, which predate Stratum V, the prevalent cooking pot is accompanied by a variety of shallow, hemispherical bowls,

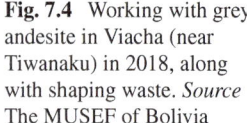

Fig. 7.4 Working with grey andesite in Viacha (near Tiwanaku) in 2018, along with shaping waste. *Source* The MUSEF of Bolivia

some of which have incised rims. These bowls likely were used as serving vessels for eating and drinking. However, these bowls disappear with the occurrence of the sandy debris that constitutes Stratum V. In higher layers, these bowls are entirely replaced by the two most common Tiwanaku shapes, the *kero,* and the *tazón* (see Fig. 4.5).

Isbell does use the term "sandy debitage" to imply that the blocks were sawn, resulting in the production of sand as waste material. However, it is worth noting that the Tiwanaku era did not possess any saws capable of cutting andesite blocks. Consequently, it becomes evident that Stratum V does not contain any splinters or stone fragments but rather a specific sand, geologically referred to as volcanic sand.

In Chap. 4, I made a connection between the invention of "high-tech" ceramics such as the *kero* and the "redware" vase or *tazon* and the development of the Tiwanaku/Pumapunku monuments. Figure 4.5 illustrates both the ancient ceramics found in layers VI-VIII at Iwawe and the technologically advanced ceramics from Tiwanaku found in layers (IV) and (III) above Stratum V. Based on the stratigraphy and the position of Stratum V, it is reasonable to infer that the construction period of Pumapunku/Tiwanaku would likely date around AD 600, as suggested by Isbell and Burkholder (2002).

It is essential to recognize that archaeologists and anthropologists misinterpreted the discovery of volcanic sand at the site at the time. In order to fully comprehend the implications of this discovery, a deep understanding of geology, particularly regarding geopolymers, is necessary. The geologists and volcanologists consulted by Isbell did not mention that andesite sand is often found in conjunction with hard volcanic rock. The presence of such sand can be detected through the phenomenon known as "*gas pipes.*" Allow me to elaborate on this phenomenon.

7.4 Carbunculus and the Gas Pipes: The Volcanic Sand in Ancient Roman Mortar

It is crucial for the reader to understand that at the Geopolymer Institute, this type of material is not a mystery or unfamiliar in our archaeological research. We have been aware of it for over 30 years because it was one of the main components, known as pozzolana or "*harena fossicia*," in the exceptional mortars developed by the ancient Roman civilization. This volcanic sand can be found in the mortars of grand monuments in Rome, such as the Coliseum, which date back to the time of Roman emperors like Julius Caesar, Augustus, Nero, and Constantine, spanning from the second century BC to the fifth century AD. According to the Roman architect Vitruvius, the ancient Latin name for this volcanic sand was "*carbunculus*" (Davidovits 2020).

In 1994, I initiated a research program called GEOCISTEM, funded by the European Union. The Geopolymer Institute, along with partners like the BRGM (Bureau de Recherches Géologiques et Minières—in English, Bureau of Research in Geology and Mining) in France, the Geology Faculties of the Universities of Barcelona in Spain and Cagliari in Italy, and the University of Caen in France, collaborated on this project (see the caption of Fig. 7.5). Our research objective was to develop new environmentally-friendly geopolymer cement using geological materials such as volcanic tuff. To achieve this goal, we chose to utilize the same deposits that were commonly exploited during Roman antiquity, over 2000 years ago, in Italy and Spain.

This background information highlights the extensive knowledge and research that has been conducted in the field of geopolymers, drawing on ancient Roman techniques and materials. It is through these investigations that we can better understand the presence of volcanic sand and its significance in archaeological sites like Iwawe.

Indeed, the research undertaken for the GEOCISTEM program required a multidisciplinary approach, bringing together experts in geopolymers, geologists specializing in volcanic rocks, and a researcher well-versed in the historical techniques of ancient Roman building materials. Frédéric Davidovits, one of the authors, provides valuable insights in his doctoral thesis titled "Geology and construction in the *De architectura* of Vitruvius" (2007), as referenced in Davidovits (2020).

To determine the nature of the "*materia*" carbunculus, or volcanic sand, an unexpected discovery was made during the GEOCISTEM program. This discovery, although not of geological origin, was significant in shedding light on the topic. During a meeting in Cagliari, Italy, in September 1996, which involved the five geologists from partner universities in the project, a visit was made to the volcanic stone quarry of Paringianu, where a hewn stone was extracted. Locally referred to as "Paringianu tuff," this volcanic rock is formed from pyroclastic flow.

The volcanic region in the vicinity consists of ignimbrite and rhyolite. The rock itself is highly indurated, meaning it is solid, and is composed of plagioclase, potassium feldspar, pyroxene, a vitreous matrix, and montmorillonite. During the visit

Fig. 7.5 The GEOCISTEM team visited the site of Paringianu, Sardinia, Italy, on September 27, 1996. From the left: Frédéric Davidovits (Caen Univ., France), Domingo Gimeno (Barcelona Univ., Spain), Philippe Rocher (BRGM, France), Carlo Marini (Cagliari Univ., Italy), Athos Rinaldi (Laviosa, Italy), Joseph Davidovits (Géopolymère, France), Sandro Tocco (Cagliari Univ., Italy), Michel Laval (BRGM, France), Luigi Buzzi (Cementi Buzzi, Italy), Jean Claude Toussaint (E.U Commission, Brussels, Belgium), (Davidovits et al. 1997)

to the quarry, which provided valuable insights into the nature of the Carbunculus "*materia*," the volcanic materials specialists made an intriguing observation. In close proximity to the extraction of very hard cut stone, the geologists showed us an unexploited area of the quarry. The tuff in this area had the same mineralogical and chemical composition as the highly durable rock, and although it contained crystals of identical dimensions, it disintegrated into sand when touched by a fingernail or finger.

The geologists explained that during the cooling process of the volcanic stratum, which requires a slow cooling rate for proper hardening, a sudden degassing event occurred in this tuff layer. This resulted in the formation of columns through which gases escaped, preventing the stone from fully solidifying as it cooled. This fascinating observation demonstrated the connection between the two types of stone: one that cooled slowly and acquired considerable consistency and the other that underwent degassing, turning it into a softer, less durable rock. According to the geologists accompanying us at the site, this difference in tenacity between rocks of similar composition is a common phenomenon.

In Fig. 7.5, you can see the members of the GEOCISTEM project standing in front of the quarry of hard volcanic rock. Afterward, they crossed the road and witnessed the gas pipes and the replicated presence of Carbunculus.

Frédéric elaborates further, stating, "Upon closer examination of the degassing columns, it became apparent that they were vertical in orientation, creating a network of small veins that traversed the entire tufa layer from bottom to surface. These columns had a height of approximately one person and were a few centimeters wide. This geological phenomenon is commonly referred to as 'gas pipes'" (Davidovits 2007).

Based on the geological and archaeological context discussed earlier, the presence of Stratum V in the archaeological stratigraphy at the Iwawe village indicates that the ground was covered with accumulations of natural andesitic sand. It is believed that this sand was likely extracted from the Cerro Khapia volcano, located in one or more locations. Subsequently, it would have been transported to the shores of Kanamarka and then ferried across the lake on rafts. Finally, the volcanic sand would have been stockpiled in the port of Iwawe (see Fig. 7.6).

In Chap. 4, we discovered that the lake level surrounding Iwawe could fluctuate depending on the seasons. It is presumed that the activities we discussed earlier took place during the rainy season when the lake was at its highest level. This scenario allowed the rafts to reach the mainland of Iwawe easily. The andesite sand was then poured onto the ground, forming a layer several tens of centimeters thick. However, as Isbell (2013) noted, this layer is not continuous and is often interrupted by large pits that have been excavated through it. This observation suggests that the area served as a storage space, where workers would dig and transport the andesite sand to another location, such as Tiwanaku or Pumapunku.

Interestingly, the volcanic sand found in the Iwawe Stratum (V) corresponds precisely to the Carbunculus mentioned in ancient Roman texts. Geologists had

Fig. 7.6 The gas pipes in the *Carbunculus* layer, volcanic sand at Paringianu, Sardinia, Italy. Reconstitution after Davidovits

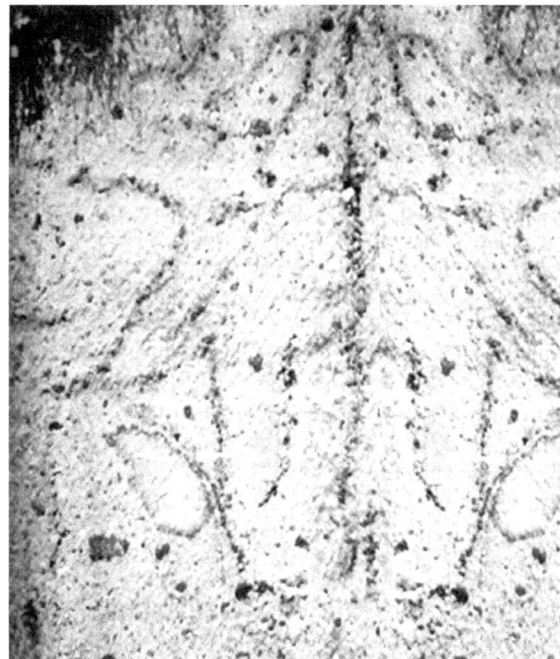

informed us that in volcanic rocks, it was not uncommon to find a combination of very cohesive stone and sand. In the case of Iwawe, the volcanic sand originates from the Cerro Khapia volcano. It is important to note that this sand is not crushed andesite stone but rather a distinct form of volcanic sand. Therefore, it stands to reason that somewhere within the Cerro Khapia volcano, there are areas where both the hard andesite rock and the sand coexist, possessing the same mineralogical and chemical properties.

The challenge, however, was determining where to search for these specific areas within the volcano. To address this, I conducted an interview with the geologists from our partner university, San Pablo, in Arequipa, Peru. Regrettably, they were unable to provide any guidance as no studies had been conducted on this particular topic. Consequently, I did not initiate an exploration program for Luis and Ralph, as I was uncertain where to send them.

It is worth emphasizing that the volcanic andesite sand found in the Iwawe Stratum (V) served as the primary raw material for the construction of the iconic "H" blocks, as well as the gates at Pumapunku and the Puerta del Sol at Tiwanaku. The utilization of this specific volcanic sand underscores its significance in the manufacturing process and highlights its role in shaping the architectural marvels of the region.

7.5   Who Transported the *Piedras Cansadas* and When?

Now, we must inquire into the investigation of why, how, when, and by whom these abandoned "piedras cansadas" (tired stones) were left along the path from Cerro Khapia to Tiwanaku/Pumapunku. To gather insights, I thoroughly examined numerous archaeological data and publications that shed light on the use of andesite rocks in Tiwanaku and Pumapunku.

Based on my research, two potential periods emerge as possible timelines for the extraction, transportation, and subsequent abandonment of these blocks along the way to Tiwanaku. The first period falls around AD 800–1000, approximately 200–300 years after the storage of andesite sand in Stratum (V) during or after the "classical" period of Tiwanaku. The second period points to the Inca Empire, around AD 1400, which would be approximately 700 years after the construction of Pumapunku.

Let us explore the first proposal. During the later stages of Tiwanaku's urban history, the Putuni platform (located at number 14 opposite Kalasasaya, as depicted in Fig. 5.17) was adorned with claddings composed entirely of andesite blocks. The architectural style of these claddings imitated the archetypes of earlier technologies. Janusek et al. (2013:74) note, "(…) *Putuni was originally built of sandstone, and that sometime later, most likely in Tiwanaku V (AD 800–1000), its more publicly visible components, notably its revetments, were rebuilt of andesite ashlars (…).*"

Although we lack written texts to provide definitive proof, anthropologists believe that after AD 800, there was a significant shift in the governance of Tiwanaku. This era saw the rise of a more authoritarian ruling class and the implementation of a policy of social division, contradicting the previous emphasis on duality and social uniformity. I previously mentioned this shift in the living conditions of Tiwanaku's inhabitants at the beginning of Chap. 5, citing Ponce Sangines (page 5): "(...) *It should be reiterated here that the type of bipartition advocated does not have implications on the content of social differentiation through classes based on economic predominance, and even less on racial consonance with individuals subjected to discrimination.*" The decline of this social tolerance may have influenced the construction methods and choice of building materials.

It is plausible that forced labor involving thousands of "second-class" individuals was employed to transport these volcanic rock blocks from Cerro Khapia. This forcing could have led to social unrest, resulting in the sacking of the temples of Pumapunku/Tiwanaku and the exile of several priests (see Chap. 10). While no definitive evidence exists, this hypothesis provides a potential explanation for the disturbances during that time.

However, the period of the Incas seems to me more appropriate because it relies on historical and archaeological facts. It is indeed known that the Incas conducted extensive restoration work at the site of Pumapunku, utilizing andesite blocks. According to American archaeologist Vranich (2013, pages 1–9), these restoration projects were substantial in nature. Vranich writes,

"*To make this history tangible, the Inca invested a huge amount of energy in constructions across the entire Titicaca Basin, modifying and enhancing important ritual places. Tiwanaku became the terminal point on an important ritual pilgrimage route that began in the Inca capital of Cuzco. The best-preserved structure at Tiwanaku, the Pumapunku, was renovated and refurbished, and a royal and religious settlement, complete with a palace, bath (essential for Inca rituals), kitchen, and associated plaza, was built against this temple. The Incas even intervened in a more aggressive fashion on the platform, emptying out fill between the retaining walls to form three chambers overlooking the plaza. Arriving in Tiwanaku, visitors and pilgrims would be hosted in the refurbished and Incanized temple, where the mythic history would be told through ritual performance.*"

It is evident that the Incas extensively employed andesite materials in their restoration efforts at Pumapunku and Tiwanaku. To acquire these materials, they likely sourced them from geological sites where naturally occurring rectangular blocks of andesite were found. This may have been the case at Kausani/Kaliri for red sandstone (as discussed in Chap. 6) and at the Cerro Khapia crater for andesite volcanics. However, it is worth noting that the Incas did not transport these large stones over the natural terrain but instead constructed carefully designed roads for this purpose. A similar road can be observed at Kanamarca, as depicted in Fig. 7.7, which connects the "quarry" of Cerro Khapia to the lake's shores (refer to the map in Fig. 7.1 for the location of the village).

Based on the information presented, it can be inferred that the Incas did not possess the knowledge of geopolymer techniques employed at Pumapunku/Tiwanaku

Fig. 7.7 **a** marked by the arrow the Inca Trail through which the andesite blocks were transported down from the crater of the volcano Cerro Khapia; **b** the "quarry." *Credit* Bardales Vassi

800 years earlier. This knowledge had either been lost following the end of the "classical" period of Tiwanaku around AD 800–1000 or was prohibited from being practiced. This lack of specialized knowledge implies a significant shift in the construction methods and materials used during these distinct periods.

We have made significant progress in identifying the geological materials utilized in the fabrication of the artificial andesite geopolymer blocks at Pumapunku/ Tiwanaku, Bolivia, around AD 600. It has been determined that these blocks were not made from crushed andesite stone but instead from natural volcanic andesite sand, which is specifically suited for this purpose. This sand possesses the same mineralogical and chemical composition as the andesite blocks found at the Cerro Khapia volcano. It is believed that this sand was extracted from one or more locations within the Cerro Khapia volcano, transported to the shores of Kanamarca in Peru, and then crossed the lake on rafts before being stockpiled at the port of Iwawe, Bolivia. At the site, the masons of Pumapunku/Tiwanaku mixed the andesite sand with an organomineral geopolymer cement to produce the andesite geopolymer blocks (Davidovits et al. 2019a, b).

Similarly, we have previously identified the geological material used in the construction of the megalithic terraces at Pumapunku. It is red sandstone that has undergone disintegration due to climatic erosion, transforming it into sandstone sand (Davidovits et al. 2019a, b). This material is associated with Kallamarka/Callamarca, a historically significant village that is part of the UNESCO World Heritage. The findings of our research have been truly remarkable, and I am proud of the work our team has accomplished thus far.

Our team planned to return to Pumapunku to verify our findings and gather representative samples. We would assess the feasibility of collecting these samples, taking into account the preservation and conservation of the site. Our commitment to thorough research and investigation remained unwavering as we strive to unravel the mysteries of Pumapunku/Tiwanaku.

Our next step was to organize a geological expedition to the Cerro Khapia volcano in the near future. This expedition would aim to discover the exact source of the andesite sand used in the construction of the monuments at Pumapunku/Tiwanaku. We hoped that these efforts would shed further light on the fascinating history and techniques of the ancient builders.

References

Davidovits J, Davidovits F (2020) Ancient geopolymers in South-American monuments, part IV: use of natural andesite volcanic sand (not crushed). Geopolymer Archaeol 36–43. https://doi.org/10.13140/RG.2.2.10021.93929/1

Davidovits J, Rocher P, Gimeno D, Marini C, Rinaldi A, Tocco S, Davidovits F (1997) Cost effective geopolymeric cements for innocuous stabilisation of toxic elements (GEOCISTEM). Final Techn Rep. Available online at: http://infoterre.brgm.fr/rapports/RR-39616-FR.pdf

Davidovits J, Huaman L, Davidovits R (2019a) Ancient geopolymer in South American monuments, SEM and petrographic evidence. Mater Lett 235:120–124. https://doi.org/10.1016/j.matlet.2018.10.033

Davidovits J, Huaman L, Davidovits R (2019b) Ancient organo-mineral geopolymer in South American monuments: organic matter in andesite stone, SEM and petrographic evidence. Ceram Int 45:7385–7389. https://doi.org/10.1016/j.ceramint.2019.01.024

Isbell WH, Burkholder IE (2002) Iwawi and Tiwanaku. In: Isbell WH, Silverman H (eds) Andean Archaeology I, chap. 7. Springer Science+Business Media, New York, pp 199–243

Isbell WH (2013) Nature of an Andean City: Tiwanaku and the production of spectacles. In: Vranish A, Stanish C (eds) Visions of Tiwanaku, Monograph 78, chap. 10. Cotsen Institute of Archaeology Press, UCLA, Los Angeles, pp 167–196

Janusek JW, Williams PR, Golitko M, Lémuz AC (2013) Building taypikala: telluric transformations in the lithic production of Tiwanaku. In: Tripcevich N, Vaughn KJ (eds) Mining and quarrying in the ancient andes. Interdisciplinary contributions to archaeology. Springer Science+Business Media, New York, pp 65–97

Mille M, Ponce Sangines C (1968) Las Andesitas de Tiwanaku. Academia Nacional de Ciencias de Bolivia Publicación No 18, La Paz

Vranich A (2013) Visions of Tiwanaku. In: Stanish C and Vranich A (eds) Visions of Tiwanaku, chap. 1, Monograph. Cotsen Institute of Archaeology Press, UCLA, Los Angeles, p 78

Chapter 8
Samples and Scientific Analysis

Samples of red sandstone—Grey andesite samples—Scientific analyses of red sandstone—Scientific analysis of grey andesite—General conclusion of the scientific analysis—The publication of scientific articles.

During the final visit to the Pumapunku and Tiwanaku sites, the team eagerly prepared to collect representative samples of red sandstone and grey andesite. It is important to note that these samples will not be randomly excavated from the monuments but carefully selected from well-referenced and non-destructive locations. Once collected, the samples from the monuments will be analyzed alongside those from the geological sites, utilizing a variety of complementary techniques. Although they are described using rock names, our investigation pointed out these materials as geopolymers. They are here referred to as using rock names for brevity.

To facilitate a better understanding of the explanations in this chapter, I will introduce some basic scientific concepts. Let us consider the case of the red sandstone found in the megalithic terraces of Pumapunku. What distinguishes natural red sandstone from geopolymer red sandstone? In natural rock, the grains of quartz and feldspar are bound together by a thin crust of clay, carbonate, or oxide minerals, which acts as cement. In geopolymer rock, the same quartz and feldspar grains are present, but the binder or cement is thicker and has a different chemical composition compared to natural cement. Geologists typically focus their analysis on the crystalline grains, which are identical in both cases, and do not typically examine the binder to determine if a rock is of the geopolymer type.

The geopolymer consists, in general, of chains or networks of mineral components containing various elements, such as silicon (Si), oxygen (O), aluminum (Al), iron (Fe), sodium (Na), potassium (K), calcium (Ca), and phosphorus (P). Geopolymer binders are created using raw materials, including alumino-silicate-based minerals like natural or calcined kaolinite clay (known as metakaolin), amorphous silica from sources like diatoms and plant silica, as well as powders from granite and volcanic rocks. Industrial by-products like coal fly ash and blast furnace slag, which are also

© The Author(s), under exclusive license to Springer Nature Switzerland AG 2024   109
J. Davidovits, *Ancient Geopolymers in South America and Easter Island*,
SpringerBriefs in Earth Sciences, https://doi.org/10.1007/978-3-031-75336-7_8

alumino-silicates, can be used as well. These minerals chemically interact through two synthesis routes currently studied in laboratories:

(1) In an alkaline medium with (Na, K) hydroxides (NaOH and KOH) and soluble alkali silicates, resulting in the formation of poly(silico-aluminates) and poly(sialates).
(2) In an acid medium, primarily using phosphoric acid mixed with organic acids (lactic acid, oxalic acid, acetic acid, citric acid, etc.).

In natural red sandstone, the cementing clay is typically kaolinite, a hydrated aluminosilicate with the formula $[Si_2O_5 \cdot Al_2(OH)_4]$. However, in the geopolymer artificial rock produced using the same raw material, the binder is a sodium-, potassium-, or calcium-based silico-aluminate with a simplified formula (–Si–O–Al–O–)–(Na, K, Ca). These binders are known as sodium, potassium, or calcium poly(sialate). Through geochemical analysis, we can differentiate between natural sandstone and geopolymer sandstone, even when the initial raw material consists of identical quartz and feldspar grains. In our analyses, we may also encounter a red geopolymer that contains a significant amount of iron (Fe) atoms, known as ferro-sialate (Fe–O–Si–O–Al–O–).

To summarize, the process of obtaining geopolymer sandstone involves taking fresh or weathered sandstone and adding kaolinite or metakaolin, which then reacts in an alkaline medium. In the case of geopolymer andesite, volcanic andesitic sand is utilized, and a geopolymer binder, which can be either alkaline or acidic, is added.

By studying the composition and properties of these materials, we aim to gain further insight into the fascinating world of geopolymer technology and its role in the construction of Pumapunku/Tiwanaku's monuments.

8.1 Samples of Red Sandstone

Upon reviewing the mineralogical and chemical study conducted by the Ponce Sangines team and published in 1971, I carefully selected two specific locations to investigate. The first spot, known as sample M2, is situated on the edge of terrace No. (1), while the second spot, M9, is located at one end of terrace No. (2). See Fig. 8.1.

One might wonder why I chose these particular sites out of the 20 others that were studied. Well, the reason is rooted in their chemical composition. Both M2 and M9 exhibit an extremely low calcium (Ca) weight content, as indicated by a reported Ca value of 0. It is this deficiency that I aim to unravel.

According to Luis, the site corresponding to sample M2, which was taken in 1970, is not ideal due to its position on the upper edge of the terrace. Additionally, it bears visible traces of inadequate and primitive excavation, as depicted in Fig. 8.2a. On the other hand, sample M9, extracted from a different section of terrace No. 2, appears to be in excellent condition and free from such markings (Fig. 8.2b).

Fig. 8.1 Location of samples M2 and M9 collected in 1970 by the Ponce Sangines team

Fig. 8.2 a Sampling location M2; **b** sampling location M9 (Ponce Sangines et al. 1971)

Fig. 8.3 Location of sample
PP-3/4, to the right of M9

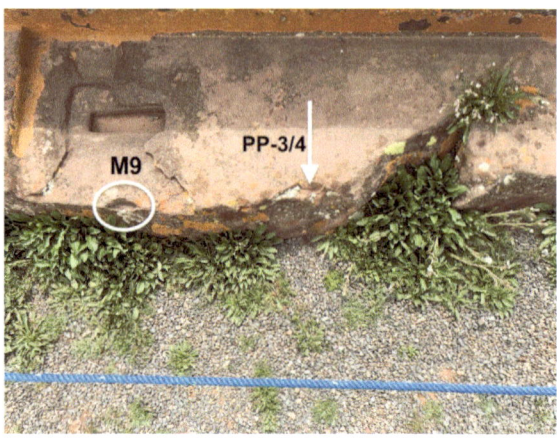

During our fieldwork, the team noticed a spot to the right of M9 that appears to have been sampled in the past. Intrigued by this, we decided to collect our sample from the same location. In Fig. 8.3, you can see this spot marked by the white arrow. I want to assure you that our sampling process does not cause any additional damage to this area. We carefully select an existing fracture that would have naturally loosened due to erosion over time.

The sample we obtain from this spot is referred to as PP-3/4, which stands for Pumapunku 3 and 4. It will undergo a comparative study with the 1970 M9 sample, as they are located in close proximity to each other.

8.2 Grey Andesite Samples

Indeed, I had not given my team any specific instructions regarding the type of andesite sample I wanted them to collect. Frankly, I had no idea where they could find any, as I had ruled out excavating the sculptures themselves. So, you can imagine my astonishment when Ralph informs me that they have managed to gather several pieces of andesite volcanic rock. He explains, *"There's a lot of debris piled up that we can easily handle."* I had actually mentioned these peculiar fragments in Chap. 4 when describing the previous phases of destruction the site had undergone. I wrote, *"Meanwhile, the H-shaped carved structural elements were too small, too hard, and made of volcanic rock resembling andesite, which is nearly impossible to cut. Moreover, their shape was unsuitable for reuse as bricks or building blocks, suggesting they were likely destroyed by a forceful impact. The team selected a piece of this particular stone during our visit (see Chapter 8)..."*. Figure 8.4 illustrates the location of sample PP1, a small andesite stone that clearly belonged to a larger structure. It exhibits two finely crafted surfaces, one perfectly horizontal and the other

Fig. 8.4 a Location of the grey andesite piece of stone chosen for sample PP1; the arrow shows the break left on the block remaining in place; **b** the two samples PP1-A and PP1-B that were kept for analysis. Look at the perfect flat surface

curved. Ralph also reports the discovery of remnants of sculpted doors or windows, which led to the collection of the small samples PP2 and PP5.

The following day, November 13, 2017, our team visited the Tiahuanaco Museum and then traveled to the city of La Paz before returning to Arequipa, Peru. To do so, they take a minibus to the border town of Desaguadero. On the Peruvian side, they manage to locate the driver who had initially brought them there.

I received in Saint-Quentin all the samples, stones, and clays that were gathered during our exploration 15 days later, on November 28, 2017. I examine them meticulously, using a magnifying glass and a digital optical microscope connected to a television screen. The following week, we will commence the SEM scanning electron microscope study and petrographic examination of thin sections. Our efforts are soon rewarded with remarkable discoveries.

8.3   Scientific Analyses of Red Sandstone

First, I examined the surface of the PP4B sample using a digital optical microscope. Following that, I turned my attention to the surfaces of the lithological samples KAU (Kausani/Kaliri), AMA (Amarillani), and MAR (KallaMarka). Figure 8.5 clearly demonstrates that the surface of KAU differs significantly from the others, particularly PP4B. This further validates the observations made at the geological sites in Chap. 6. It is now conclusive that the geological site of Kausani/Kaliri, which Bolivian and American archaeologists had highly regarded, can be definitively ruled out.

8.3.1   Thin Sections

For a thorough petrographic analysis, it is necessary to create thin sections of the samples. These thin slices of rock are stabilized with resin and require specialized equipment, such as a petrographic microscope with polarizing light. Since we lack this equipment, we entrusted this study to the Geosciences Laboratory at the Institut Polytechnique UniLaSalle de Beauvais, a specialized university laboratory located in our region. Frédéric took our samples to Beauvais, and I received their initial report in April 2018.

The report revealed a significant difference between the monument sample PP4 and the others. With the report in hand, I sent it to Luis, who happened to be in Arequipa, Peru. Luis confirmed that PP4 contains an excessive amount of binder, or what geologists refer to as "cement." This finding reinforces the artificial nature

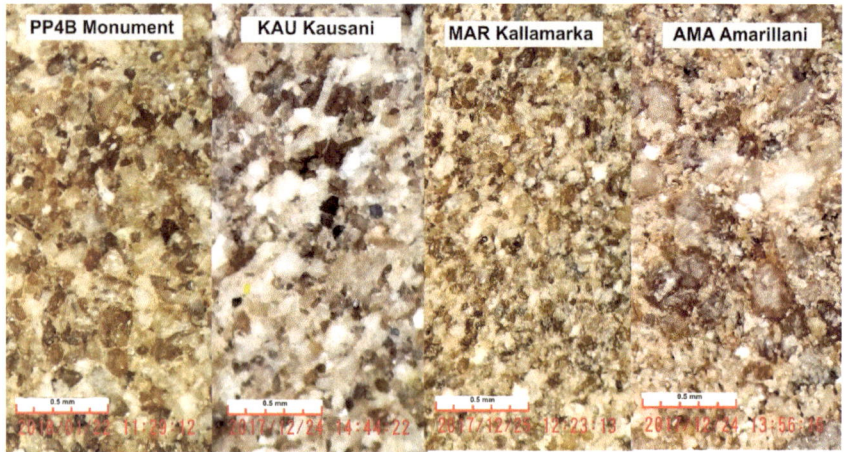

Fig. 8.5   The surface of the different samples of red sandstone, PP4B (Pumapunku monument), KAU (Kausani), MAR (Kallamarka), AMA (Amarillani). The scale is 0.5 mm

of this particular type of red sandstone. In natural sandstones like KAU, AMA, or MAR, the grains of quartz and feldspar are typically coated with a thin crust of clay. Occasionally, small lumps of clay can be found scattered throughout (marked as C). These reddish accumulations are clearly depicted in Fig. 8.6 for MAR-2. However, in PP4-1 and PP4-2, the grains are coated with remarkably thick and wide covers of a red binder or cement, which Luis describes as highly fluid, akin to liquid cement.

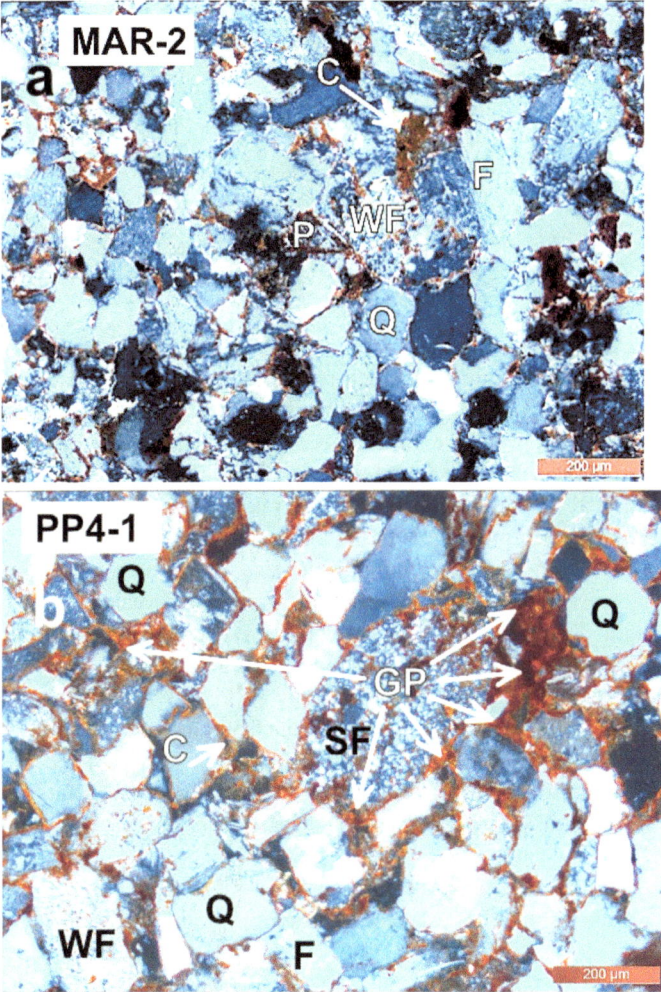

Fig. 8.6 Thin sections of red sandstone: **a** MAR-2 (Kallamarka); **b** PP4-1 (Pumapunku); **c** PP4-2 (Pumapunku). *Q* quartz, *F* feldspar, *WF* weathered feldspar, eroded feldspar, *VC* volcanic grain, *C* clay, *P* plagioclase, *SF* stone fragments, *GP* geopolymer, *Scale* 200 μm. Transmission polarised light analysis

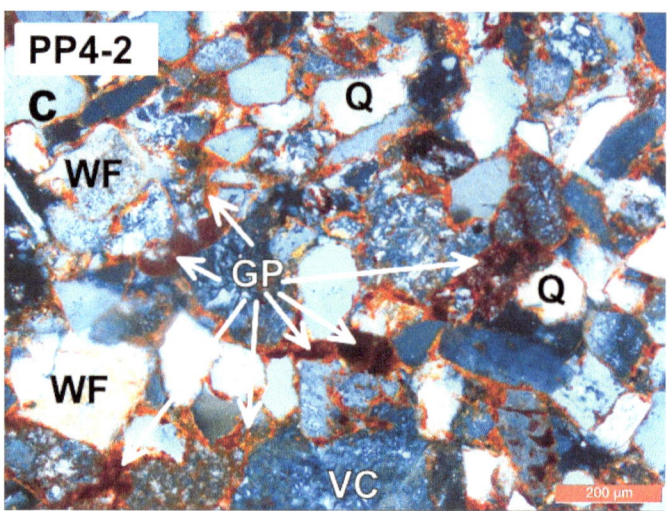

Fig. 8.6 (continued)

These cement-rich features reminded me of the conclusion I had drawn when studying the chemical analyses conducted by Bolivian scientists in 1970. In Chap. 6, they revealed that there is five times more binder (cement) in the Pumapunku sandstone than in the sandstone found at Kausani/Kaliri.

I also requested the Geosciences Laboratory to conduct X-ray diffraction analyses, as it provides information about the clay content in the sandstone. For KAU, AMA, and PP4, no clay was detected, or if present, the quantity was too small to be quantified. However, for MAR, the X-ray diffractogram revealed a clay content of 7.7% by weight. This concentration corresponds to the several small lumps of clay marked as C in the MAR-2 thin section in Fig. 8.6.

It is worth noting that the absence of clay in PP4 is intriguing and warrants reflection on my part. When we submitted our samples for analysis, we did not specify that we considered PP4 to be an artificial material. After receiving all the results, I sought clarification from the laboratory regarding a few points, specifically asking, "*If the thick, fluid binder observed in PP4 is not clay, what is its mineralogical composition?*".

The response I received was as follows: "*It is a cement composed of argillite and iron oxides.*" This information was recorded in Photo 13 of the report, reproduced in Fig. 8.7.

It is important to note that the clay mentioned has not been misidentified as argillite. Clay is a well-crystallized alumina silicate that can be easily detected by X-ray diffractometry, whereas argillite is a mixture of undefined and X-ray amorphous (often undetectable) aluminosilicates. In essence, the interpretation presented in Fig. 8.7 reflects the definition provided by someone unfamiliar with the scientific

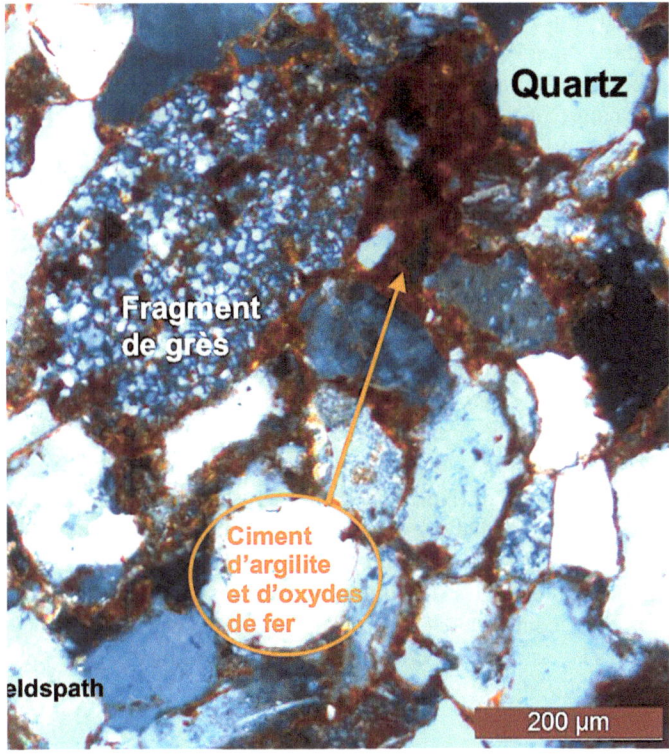

Fig. 8.7 Photograph No. 13 of the UniLaSalle report with the words "*Cement made of argillite and iron oxides*" (in French: ciment d'argilite et d'oxyd es de fer) and "*Sandstone fragment*" (in French: Fragment de grès)

terminology, which aligns with what we refer to as a "ferro-sialate or ferro-silico-aluminate" geopolymer. However, geopolymers are also X-ray amorphous, like many argillites, making them undetectable through X-ray diffractometry analysis. Unbeknownst to the geologist, they inadvertently confirmed the artificial nature of this PP4 sandstone.

8.3.2 Scanning Electron Microscope (SEM) Analysis

On January 16, 2018, I began examining the Pumapunku stone samples using our scanning electron microscope (SEM). To ensure proper sample preparation, I enlisted the help of Mathilde Maléchaux, the laboratory technician responsible for physical measurements of geopolymeric materials. This laboratory is located at the premises of Pyromeral Systems, which is owned by my brother Michel and his son Jean-Michel, near Senlis, north of Paris (as mentioned at the beginning of Chap. 1).

As I mentioned in Chap. 1, SEM analysis not only provides visual images but also reveals the atomic composition of the elements through energy-dispersive X-ray spectroscopy (EDS) analysis. This analysis will play a crucial role in uncovering the true nature of the geopolymer binder, which I refer to as the "geological glue." According to the legends shared by Peruvian ethnologist Francisco Aliaga (as discussed in Chap. 1), this binder was believed to have been derived from plant extracts. Therefore, one of our primary objectives is to search for the signature of carbon atoms (C) within the sample. However, to my slight disappointment, no carbon atoms were detected. It appears that the geological glue is entirely mineral-based, specifically classified as a ferro-sialate geopolymer (Davidovits et al. 2019a).

In this chapter, I will focus on the essential results, precisely the nature of the ferro-sialate geopolymer binder and the sodium (Na) atom content. Figure 8.8 showcases the geometric structures that have crystallized over several hundred years, from the time of the monument's construction until today. The EDS spectrum reveals the following amounts (atoms %) for the major elements: 59,72 silicon (Si), 15,43 aluminum (Al), 7.63 sodium (Na), 3.70 potassium (K), 1.87 magnesium (Mg), and 11.65 iron (Fe), confirming the presence of a ferro-sialate geopolymer binder. In this case, iron substitutes for aluminum, resulting in a Si/Al + Fe ratio of 2.3. The carbon (C) is insignificant. It is part of the equipment.

The sodium (Na) content was only marginally determined by the Ponce Sangines team in 1970–1971, as they did not have access to the advanced equipment, such as the scanning electron microscope, that we have today. Here are the sodium (Na) values obtained:

- KAU (Kausani): 6.67%
- AMA (Amarillani): 1.56%
- MAR (Kallamarka): 5.10%
- PP4 overall: 7.63–10%.

Fig. 8.8 SEM backscattering image of ferro-sialate matrix and its geometric structures (arrows), on the right, the EDS spectrum   .

The sodium (Na) content of PP4 is significantly higher than that of KAU, AMA, and MAR. Therefore, if we assume that Pumapunku PP4 is a natural sandstone, it does not belong to the sandstone found in the Quimsachata mountain range south of Tiwanaku.

8.3.3 Conclusion for Red Sandstone

Indeed, none of the analyses conducted on the geological samples in 1970 revealed such high levels of sodium (Na). This type of sandstone has not been found in the region, raising questions about its origin. If we adhere to the traditional argument that the monument is constructed from natural sandstone, then this sedimentary rock does not belong to the local area. Consequently, according to conventional archaeology, the massive megalithic slabs weighing between 100 and 180 tons must have been extracted and transported from a geological site located elsewhere, potentially far away. The transportation of these colossal sandstone blocks, equivalent in size to a house (8 m × 8 m), would have required primitive sleds over treacherous terrain, as depicted in Fig. 6.5 in Chap. 6. This notion is challenging to accept, although archaeologists have experimented with dragging smaller pillars (1 to 5 tons) on flat ground.

However, the X-ray study revealed the presence of kaolinite clay in the MAR (Kallamarka) sample. Therefore, if we entertain the idea that the MAR Kallamarka site is the source of the monumental sandstone, it would necessitate the addition of an alkaline hardener containing sodium (Na). This mixture would create an alkaline geopolymer by reacting with the kaolinite clay. Consequently, we are witnessing a remarkable technological innovation that arises as a direct consequence of the development of "high-tech" ceramics in Tiwanaku, employing the LTGS technology elucidated in Chap. 4, Sect. 4.2.2.

The skilled stonemasons could have procured natron salt and sodium carbonate together with calcite from Laguna Cachi as potential resources. This small lake, situated south of the vast Salar de Uyuni in the Altiplano, could have served as a viable source. Caravans were known to travel between Tiwanaku and the southern regions on a regular basis, maintaining a constant flow of goods. The calcite could have been calcined in a primitive ceramic kiln at temperatures of 650 °C, as described in Chap. 4.

In Fig. 8.9, we find ourselves at an elevation of 3800 m above sea level, with San Pedro de Atacama located 750 km away from Tiwanaku. According to archaeological records, the llama caravans would have passed through Laguna Cachi after traversing the expansive Salar Uyuni. It is conceivable that the ancient builders of Pumapunku/ Tiwanaku 1500 years ago may have exploited the natron salt sodium carbonate from this area. Remarkably, evidence suggests that the extraction of natron salt from Laguna Cachi has persisted even into our modern era, spanning several centuries.

Fig. 8.9 Laguna Cachi and
the path of the llama
caravans linking Tiwanaku to
San Pedro de Atacama
(Google Earth)

8.4 Scientific Analysis of Grey Andesite

Upon examining the PP1 andesite sample, I utilized the digital optical microscope right from the start. The photograph in Fig. 8.10 was taken on December 29, 2017, capturing the surface in all its elegance. To my astonishment, I beheld a sight of exquisite beauty adorned with remarkable crystals. What struck me as peculiar was the absence of any previous examination of these unique "H" blocks under an optical microscope. It is pretty strange, really. Some individuals attribute these structures to the work of aliens or an advanced civilization, yet none of them took the simple step of scrutinizing it with an optical microscope. Even to this day, I have not come across any publication, article, book, or video that documents such information.

As I observe the surface, I see plagioclase crystals in a radiant white hue (#1), along with elongated crystals in darker shades. These constituents align with the typical characteristics of andesite volcanic rocks, such as amphibole, pyroxene, and biotite (#2, #3). The surface appears remarkably flat, devoid of any signs of polishing caused by abrasive grains or evidence of tool marks. However, it is punctuated by small holes ranging from 0.2 to 0.5 mm in depth, exhibiting sharp edges (#4, #5 and #6). At the base of these cavities, I can discern the presence of andesite minerals, confirming the composition of the stone.

Fig. 8.10 Surface of grey
andesite PP1, digital optical
microscopy. The numbers 1
to 6 designate the crystals
and holes explored. The
scale is 1 mm

8.4.1   Thin Section

The UniLaSalle laboratory also conducted an examination of a grey andesite sample,
along with the red sandstone. This particular investigation focused on sample PP2, a
small fragment measuring just a few centimeters, which was extracted from carved
debris discovered near PP1. The results of this analysis can be found at the beginning
of the UniLaSalle report, accompanied by detailed photographs of thin sections that
provide valuable insights. One of the images, Fig. 8.11, displays a thin section that
showcases the intricate details of the sample. In this portion, we can observe the
presence of tiny plagioclase crystals, depicted in white, as well as larger pyroxene
crystals. Additionally, scattered throughout the thin section, there are distinct areas
of black amorphous substance.

Interestingly, similar mentions of this amorphous substance were made by 19th-
century travelers like Alphons Stüble and Max Uhle (1892) in their own thin sections
cut from andesite samples taken from a monument in Tiwanaku. It is important to
note that their findings do not differ from our PP2 sample. In German, they describe it
as "*Runde Nester amorpher Substanz, in Mitte licht braun gefärbt, nach den Rändern
verblassend, wurden vereinzelt bemerkt*," which translates to "Round nests (pockets)
of amorphous substance, light brown in the center, fading at the edges, have been
noticed from time to time (individually)."

8.4.2   Scanning Electron Microscope Analysis

Now, let us shift our attention to hole number 4, marked in Fig. 8.10, which we
will examine with a higher magnification using the optical microscope and, later, the
scanning electron microscope.

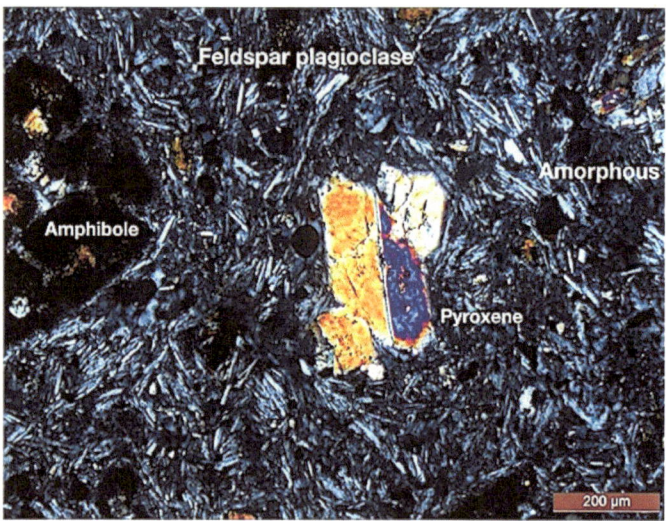

Fig. 8.11 Thin section of the PP2 grey andesite sample, transmitted polarised light: tiny plagioclase crystals, amphibole and pyroxene crystals, amorphous material

Within the depths of this particular hole, as depicted in Fig. 8.12, on the right, we encounter a composition of black matter in the middle of various minerals, encompassed by white feldspar crystals on the surface. The corresponding SEM image in Fig. 8.12, on the left, unveils plagioclase (Plag) crystals on the surface, while within the hole, we observe clusters of polycrystals such as hornblende (H), pyroxene-augite (P-A), and a ferro-silicate (Fe-Si) inclusion. Interestingly, lying between these crystals, denoted by a white square, we encounter a substance that does not align with any known classified mineral when observed through the optical microscope.

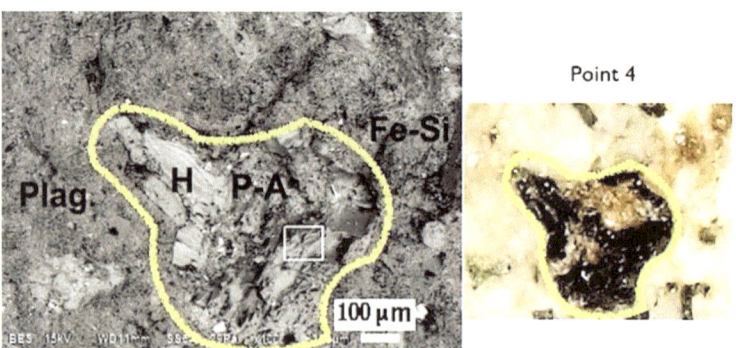

Fig. 8.12 Right, point 4 in Fig. 8.10 with optical magnification; left: SEM image of the same point 4 with *Plag* plagioclase, *H* hornblende, *PA* pyroxene-augite, *Fe-Si* ferro-silicate

Upon closer examination, as seen in Fig. 8.13 under high magnification, we stumble upon a surprising element—completely amorphous in nature, resembling resin rather than a crystalline mineral. Could this be the same amorphous material mentioned earlier in the thin section of Fig. 8.11? It was through this investigation with the SEM that I started Chap. 1 of this book. The analysis conducted using energy-dispersive X-ray spectroscopy (EDS) revealed the presence of a significant amount of carbon (C) and nitrogen (N) alongside other mineral elements such as sodium (Na), magnesium (Mg), aluminum (Al), silicon (Si), phosphorus (P), sulfur (S), chlorine (Cl), potassium (K), and calcium (Ca). The nitrogen (N) peak, although relatively small, is situated on the left side of the spectrum between the prominent carbon (C) and oxygen (O) peaks. Due to the limitations of the EDS analysis, the concentration of nitrogen (N), being a light element, cannot be precisely determined. However, I can confirm its qualitative presence in relatively high quantities within this amorphous organo-mineral substance, possibly indicating the presence of an organic ammonium-based compound.

The discovery of biological organic matter within volcanic rock is genuinely remarkable. Some individuals may argue that carbon (C) and nitrogen (N) are also components of gases that could have been trapped within the volcanic rock, such as carbon dioxide and nitrogen from the air. However, I would like to emphasize that these arguments stem from a lack of understanding regarding the operation of electron microscopes. These instruments require a highly controlled vacuum environment, often necessitating the running of pumps for extended periods, sometimes up to 10 to 20 min or more, prior to the commencement of the study. This setting ensures the evacuation of all gases present, as any remaining gases would interfere with the functionality of the device. Therefore, any carbon and nitrogen atoms detected through EDS analysis belong to solid substances, such as organic molecules. This revelation is highly unusual and challenges our understanding of natural processes, as there is no known organic matter capable of withstanding the extreme temperatures of magma and erupting lava, specifically within andesite volcanic stone. From these findings, we can only deduce that this particular sample is artificial.

8.4.3 Conclusion for Grey Andesite

Indeed, an argument could be raised, suggesting that the SEM image we examined, taken from a hole on the surface of sample PP1, might be influenced by surface contamination. To address this concern, we conducted further investigations within the interior of PP1 by extracting a smaller sample known as PP1C. By analyzing spots located one centimeter and two centimeters below the surface, we discovered several instances of the same black organic matter exhibiting identical EDS spectra. For a comprehensive understanding of these findings, interested readers can refer to the scientific publications, articles available on the internet, and accompanying videos (Davidovits et al. 2019b, c).

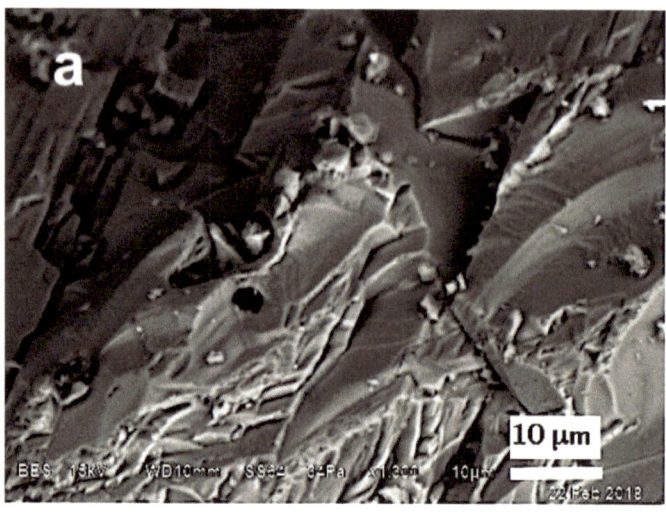

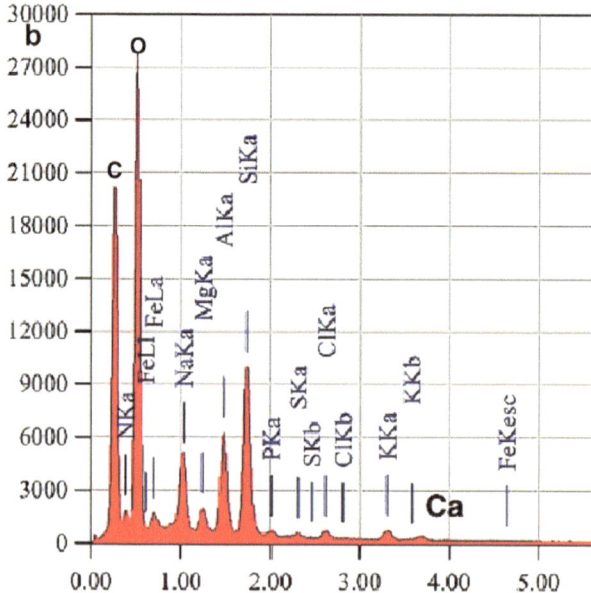

Fig. 8.13 a SEM backscattering image of amorphous material of point (4), white square in Fig. 8.12. It looks like resin, scale 10 microns. The date is February 22, 2018; **b** EDS spectrum

The presence of biological organic matter consistently observed throughout the sample effectively eliminates the possibility of contamination. It becomes increasingly apparent that this biological organic matter is indicative of the presence of an artificial stone. This significant evidence aligns with the notion that the sample we have examined is indeed man-made.

8.5   General Conclusion of the Scientific Analysis

The examination of a thin section from the red sandstone monument in Pumapunku reveals an intriguing feature. The sandstone grains are coated with a thick, fluid red ferro-sialate geopolymer matrix, a characteristic that is highly uncommon in naturally formed sandstone. This observation strongly supports the concept of artificial sandstone geopolymer concrete. Further analysis using SEM/EDS techniques to detect elements like Na, Mg, Al, Si, K, Ca, and Fe suggests that the Kallamarka site could be the potential source of the megalithic blocks found at Pumapunku. These colossal slabs, weighing between 100 and 180 tonnes, were crafted approximately 1500 years ago. It is speculated that the builders of these structures may have transported geologically degraded sandstone to create their geopolymer sandstone concrete.

Typically, sandstone consists of quartz and feldspar grains. However, due to climatic erosion, feldspar undergoes degradation and transforms into clay minerals, mainly kaolinite. This process weakens the rock, causing it to disintegrate. This loose rock has been discovered in Kallamarka and is presented and explained by our team geologist, Luis, in Chap. 6. Interestingly, this degraded sandstone serves as an excellent raw material for producing geopolymer sandstone. To complement the mixture, the Pumapunku masons likely incorporated lime and metakaolin, produced in a ceramic kiln, along with alkaline reactive elements like natron (Na_2CO_3) extracted from Laguna Cachi, a small lake located south of the grand Salar de Uyuni in the Altiplano region of Bolivia.

However, the Pumapunku site's most controversial aspect lies in the smaller and more puzzling objects known as the "H" sculptures, which are crafted from andesite volcanic stone. The SEM study conducted on this artificial grey andesite reveals the presence of biological organic matter, potentially the raw material for the geopolymer binder. The analysis of this material yields carbon, nitrogen, and various minerals. The existence of amorphous biological organic matter within volcanic stone is highly unusual, if not seemingly impossible. This discovery has also been detected in thin sections examined through optical and SEM studies, further strengthening the idea of artificial andesite geopolymer concrete.

The culmination of these analyses raises an important question: What chemistry was employed in the creation of these structures? In this reconstituted andesite, we do not find polysialate- or ferro-sialate-based geopolymer as observed in the red sandstone megaliths, indicating the absence of an alkaline medium. Drawing from my knowledge of geopolymer science, I am led to believe that if an alkaline medium is not present, an acidic one must have been utilized. This possibility aligns with ancient legends reported to me by Francisco Aliaga, often overlooked by traditional archaeology, that mention a substance of plant origin capable of softening stones. These accounts correspond to the potential use of plant extracts containing organic acids to fulfill this function. Such discussions are prevalent in South America, harkening back to the themes explored in Chap. 1.

To create geopolymer-andesite concrete, the skilled masons of Pumapunku likely transported a volcanic material with a sandy consistency. This specific material was discovered and discussed in Chap. 7, particularly in relation to the "Stratum V" layer found during the archaeological exploration of Iwawe on the Bolivian shores of Lake Titicaca. The andesite sand originates from the Cerro Khapia site, situated on the Peruvian side of the lake. This sand would have been combined with an organo-mineral geopolymer binder made from local biomass ingredients. Our research indicates that these architectural components may have been shaped using a wet sand geopolymer molding technique. Alternatively, it is possible that a preform was created through molding. Then, while the geopolymer was still soft and malleable, before complete hardening, it would have been sculpted using the tools available during that era, such as wood, stone, or obsidian.

One aspect that continues to surprise me in the results of our analyses is the presence and utilization of two distinct geopolymer technologies. The first technology, employed in the creation of red sandstone megaliths, utilizes ingredients sourced from the mineral world, such as sandstone, clay, and natron salt. This process involves an alkaline medium and results in the formation of a mineral geopolymer of the ferro-sialate type. The second technology, employed in the construction of the grey andesite gateways and other "H" sculptures, is markedly different. It relies on plant extracts and biomass, representing the life-giving world, to produce an organo-mineral geopolymer.

This duality between the mineral world and the life-giving world is beautifully captured in the words of American anthropologist Janusek (2008, p 134): "(…) These red sandstone constructions conjured the memory of ancient deeds and beings …. and invoked the ancient bedrock of the local mountains. (…) The bluish-grey andesite evoked the life-giving principles of more distant sacred volcanic peaks venerated by numerous altiplano communities while symbolizing the complementary life-principle of water in Lake Titicaca (…)."

8.6   The Publication of Scientific Articles

In order to appreciate the duality of the geopolymer technologies, we decided to publish our findings accordingly. I began by writing the first scientific article focused on the artificial nature of the red sandstone megaliths, representing the mineral world. On April 23, 2018, I submitted my manuscript to the chief editor of the journal "*Materials Letters*," which is well-known in the scientific community. After a thorough peer-review process, I received the reviewer's report on August 13. We made the necessary revisions and sent back the revised manuscript on August 20. Following a second round of peer review, our article was finally accepted on October 6, 2018. It took over six months for the publication process, but the outcome was worth it. The title of this paper is "*Ancient geopolymer in South American monuments: SEM and petrographic evidence*," authored by Davidovits et al. (2019a).

With the first article successfully published, I then turned my attention to writing the second paper, dedicated to the geopolymer of the life-giving world, the grey andesite. Since "*Materials Letters*" did not permit publication on this topic due to their editorial arrangements, we sought another renowned international journal, "*Ceramics International*." On December 17, 2018, we submitted my second article to this journal. The fact that my first paper had been accepted by "*Materials Letters*" worked in our favor, and the editor received the slightly revised manuscript on January 3, 2019. To my surprise, the article was accepted on the same day. The title of this publication is "*Ancient organo-mineral geopolymer in South-American Monuments: Organic matter in andesite stone. SEM and petrographic evidence*" written by Davidovits et al. (2019b). The rapid editorial process, taking only 15 days, generated a significant buzz on the internet and social networks, particularly in Latin America.

Now, I can finally shed light on the chemical ingredients involved in the production of the geopolymer technology based on the life-giving world. It is time to unveil my discoveries and delve deeper into these questions in the next chapter.

References

Davidovits J, Huaman L, Davidovits R (2019a) Ancient geopolymer in South American monuments, SEM and petrographic evidence. Mater Lett 235:120–124. https://doi.org/10.1016/j.matlet.2018.10.033

Davidovits J, Huaman L, Davidovits R (2019b) Ancient organo-mineral geopolymer in South American monuments: organic matter in andesite stone, SEM and petrographic evidence. Ceram Int 45:7385–7389. https://doi.org/10.1016/j.ceramint.2019.01.024

Davidovits J, Huaman L, Davidovits R (2019c) Tiahuanaco monuments (Tiwanaku/Pumapunku) in Bolivia are made of geopolymer artificial stones created 1400 years ago. Archaeological Paper #K-eng, Geopolymer Institute Library, http://www.geopolymer.org. https://doi.org/10.13140/RG.2.2.31223.16800

Janusek JW (2008) Ancient Tiwanaku. Cambridge University Press, New York. ISBN: 978-0-521-01662-9

Ponce Sangines C, Castaños EA, Avila SW, Urquidi BF (1971) Procedencia de las Areniscas utilizadas en el Templo Precolombino de Pumapunku (Tiwanaku). Academia Nacional de Ciencias de Bolivia, Publication No 22, La Paz

Chapter 9
How Did I Discover the Geopolymer of the Life-Giving World?

Some traces in North America—Why guano?—Archaeological documents—The Guano samples from the Punta Coles reserve in Ilo—Analysing the Guano from Ilo.

On September 20, 1982, I was in southern Arizona in the USA, contemplating the peculiarities of a strange archaeological site, the Casa Grande Ruins National Monument (Fig. 9.1), on a trip with my son Marc. It was a fascinating experience to observe the unique characteristics of that archaeological site. The presence of giant cacti in the area, with their ability to produce large quantities of juice containing organic acids like citric acid, serves as a perfect illustration of the point I made in our paper presented at the congress in Bradford, England, earlier that year (1982).

9.1 Some Traces in North America

It is indeed intriguing to learn about the Casa Grande Ruins National Monument and the ancient Hohokam people who constructed these structures during the early thirteenth century. The name "Casa Grande," meaning "big house" in Spanish, refers to the largest remaining four-story structure on the site. Despite the passage of seven centuries and exposure to extreme weather conditions, this caliche structure made of sedimentary rock composed mainly of limestone binder has managed to endure.

While traditional adobe processes were used in the construction of these structures, it is important to note that the material utilized is not sun-dried clay but rather caliche. Caliche is the opposite of clay and does not possess the properties necessary to shape a paste or get wet. This raises the question of how the ancient inhabitants were able to work caliche in their construction methods.

As I mentioned earlier, one possible explanation lies in the incorporation of organic acids to create a limestone paste, similar to clay. It is conceivable that the masons of that time utilized the juice from cacti to prepare a paste that, upon drying, would

J. Davidovits, *Ancient Geopolymers in South America and Easter Island*,
SpringerBriefs in Earth Sciences, https://doi.org/10.1007/978-3-031-75336-7_9

Fig. 9.1 **a** Joseph in front of the Casa Grande Ruins; **b** Marc and the giant cactuses in the rear of the building protected by a roof; **c** the map shows the location of the Casa Grande Ruins between Phoenix and Tucson (it is not the city Casa Grande)

Fig. 9.1 (continued)

become a material capable of withstanding the harsh climate for a certain period. This could explain why the structures have persevered for over 600 years. However, it is essential to note that the caliche and organic acid mixture alone does not possess the enduring qualities necessary to create the weather-resistant andesite found in Pumapunku. The secret behind the creation of Pumapunku's man-made "andesite" lies in the addition of a hardener that chemically transforms the material into a long-lasting rock capable of braving the elements.

It is fascinating to see how the geopolymer technology I have been exploring extends beyond the Andes and into southern Arizona. Sadly, the indigenous people in that region only retained knowledge of the dissolution and disintegration phase of the process, lacking crucial information on how to harden the material again. This left me in a situation similar to when I wrote about it in Chap. 1, pondering the question of how to transform the paste into a durable, weather-resistant rock. It was clear that more research and testing were needed to find the answer.

Over the years, I received various messages suggesting plant extracts that could potentially dissolve stones. However, most of these suggestions were simply reiterations of information found on the internet without providing a solution for the hardening process. There were experiments with the Jotcha shrub extract by Father Lira, which resulted in a paste that could not be hardened back into stone. One intriguing story I came across was that of Percy Fawcett, an explorer from the early twentieth century, who mentioned plants in the rainforest capable of disintegrating shoe soles and certain metals. Recommendations also came in for the use of fluoroacetic acid, a toxic mineral-organic acid found in various plants, but it posed significant safety concerns due to its corrosiveness.

All of these proposals focused on substances that had destructive properties—dissolving, disintegrating, or cutting. None offered a solution for hardening the soft pastes, reconstitution, or creation. We did not need these dangerous and corrosive chemicals. The answer was much more straightforward: vinegar. Acetic acid, easily obtained from plant extracts and juices containing sugar, such as the renowned chicha of the Tiwanaku people, provided the necessary organic acids. Citric acid from cacti and other food plants also served as additional sources. After publishing our analysis results, revealing the artificial nature of the andesite samples and the presence of organic matter, I gave several lectures on the topic. It was during these discussions that I introduced the idea of the hardener: guano.

In my research, I discovered that guano, which is still commonly used as fertilizer, had been transported by llama caravans from Ilo on the Pacific Ocean to Tiwanaku via Moquegua. This region is depicted on the map in Fig. 9.3. While guano is indeed an excellent fertilizer, its transportation to the Altiplano was not solely for that purpose. With all the information I could not share during those lectures, I proposed that guano was the missing ingredient—a key component in the hardening process that transformed the soft paste into a durable rock.

9.2 Why Guano?

To fully grasp the origins of this seemingly absurd idea in archaeology, we must revisit Ralph's expedition to Easter Island in 2016, which I detailed in Chap. 1. I prepared a study program for him and advised him against wasting time visiting Orongo, which is southwest of the island, where the Bird Man cult was practiced. It appeared to be a recent cult with no apparent connection to the other monuments on the island. However, Ralph had ample time and managed to capture some photographs. The Bird Man ritual involved a competition to retrieve the first *manutara* egg, which belonged to the grey-backed stern or sooty stern seabirds that regularly nested on the islets of Easter Island.

This ritual emerged after the decline of the Moais (the statues) around 1550 and continued until 1820. The contest took place on the neighboring islet, Motu Nui, located 1.5 km away in the ocean (refer to Fig. 9.2). Representatives from each clan awaited the arrival of the first stern eggs of the season. The individual who secured the initial egg would swim back to Easter Island and, upon scaling the cliffs of Orongo, present the egg to their sponsor in front of the judges of Orongo. The victor's sponsor would then be bestowed with the title of Tangata Manu, granting them significant power on the island for the year.

After Ralph's return, a few months passed, and I found myself immersed in various documents about Easter Island. These writings emphasized the presence of guano on the islets of Motu Nui, Motu Iti, and Motu Kao Kao, as well as one near the Poiké trench in the east called Motu Marotiri. It was at this point that an intriguing question arose within me: could the cult of the stern's egg be a remnant of a time when the people of Easter Island utilized guano to harden their stone? Considering my knowledge of guano's chemical composition, I immediately grasped the implications of this idea. It was indeed plausible that mixtures of stone paste and organic acids derived from plant extracts could be hardened using guano.

This revelation led to a series of subsequent inquiries. Could guano be connected to Tiwanaku? Once again, the answer was affirmative. I stumbled upon documents discussing the renowned Peruvian guano that was extensively exploited during the nineteenth century and continues to be used today (refer to Table 9.1). Additionally,

Fig. 9.2 Guano-covered islets on Easter Island; Motu Kao Kao in the foreground, Motu Iti in second, and Motu Nui in third, seen from the Orongo cliff

Fig. 9.3 Guano trade around AD 600 between Ilo and Tiwanaku, after Minkes (2005)

I came across studies written by anthropologists describing the transportation of this very same guano, around AD 600, twelve centuries earlier, from the Pacific Ocean up to the Altiplano, specifically to Tiwanaku.

Table 9.1 Chemical composition of three Peruvian guanos containing mainly ammonium oxalate and urate, calcium oxalate, ammonium phosphate, and calcium phosphate, according to Towers (1845)

Uric acid, ammonium urate	17.92	15.27	7.80
Ammonium oxalate	7.40	3.45	7.50
Ammonium phosphate	8.80	11.10	–
Organic matter (oxalate?)	8.76	41.73	24.00
Calcium phosphate	22.00	10.25	29.30
Calcium Oxalate	2.50	–	–
Kalium sulphate + sodium chloride	8.00	12.90	11.40
Water	22.00	6.50	20.00

9.3   Archaeological Documents

Archaeology has provided us with valuable insights through various texts written during the Spanish conquest, including Pedro Cieza de León's Chronicle of Peru, which I discussed in Chap. 4. These texts transcribe the oral accounts shared by the indigenous people of that time. Among these texts, there is one that sheds light on the trade of guano between Ilo on the Pacific Ocean and Tiwanaku, spanning a journey from sea level to an impressive altitude of 3800 m above sea level (as shown in Fig. 9.3). This particular text was examined by W. Minkes in his research titled "Wrap the Dead," published in the Archaeological Studies of Leiden University (The Netherlands) in 2005. The relevant Chaps. 5.5.2 and 6.5.2.

In Minkes' study, he highlights the significance of the Ilo site on the Pacific Ocean and a site known as "El Descanso" or "Place of Rest" in Spanish. This name has been passed down orally and refers to the historical use of the site as a resting place for llama caravans traveling to and from the highlands via Moquegua. Historical documents indicate that the Moquegua Valley served as a popular route for these caravans, carrying large quantities of guano harvested from Punta Coles, Ilo, to Tiwanaku. This guano trade appears to have been particularly active during the construction of Tiwanaku and Pumapunku around AD 600, perhaps driven by the growing demand for guano during that time. In exchange for the guano, the coastal population of Ilo received coca, camelid wool, dried meat, and llamas to assist in transporting the valuable ingredient.

These findings provide us with valuable historical and archaeological evidence of the trade routes and exchanges that took place between different regions, contributing to our understanding of the cultural and economic dynamics of the time.

While guano indeed serves as an excellent fertilizer, I respectfully disagree with the opinion of some archaeologists regarding its primary purpose for being transported to the Altiplano. The Tiwanaku civilization had already developed a unique agricultural system known as the "raised-field system" prior to the exploitation of guano (Kolata and Ortloff 1989). This innovative farming method involved the use of raised, elongated planting beds surrounded by water-filled ditches. These ditches contained aquatic plankton and small fish, which naturally fertilized the fields. Consequently, the Tiwanaku people did not rely on guano as they were able to produce their own organic manure on-site. Therefore, it would be inaccurate to state that guano was sent to the highlands solely for agricultural purposes, as this civilization had already established self-sufficient agricultural practices.

Furthermore, I would like to provide some information regarding the composition of Peruvian guano, as indicated in Table 9.1. These analyses were conducted around 170 years ago on guano samples imported into Europe, specifically England and France. It is worth noting that the analyses vary significantly due to the volatility of certain elements, such as water and ammonia, which tend to dissipate over time. The composition of guano primarily consists of various salts of organic acids, including oxalates and phosphates, as well as an undefined organic matter that may be oxalate-related.

In my research, I have come to the conviction that guano was not primarily used for agriculture, considering the substantial quantities extracted that surpass the needs of mere farming. Instead, I propose that guano served as an organic hardener in the creation of the geopolymer technology that shaped their world. This theory is supported by the presence of chemical ingredients in guano, such as calcium phosphate, ammonium oxalate, and calcium oxalate, which, when combined with one or more organic acids, can transform into phosphoric acid and oxalic acid. These acids play a crucial role in the hardening process of the mixture. For more detailed information, I refer you to the scientific publications and videos cited in the accompanying references.

9.4 The Guano Samples from the Punta Coles Reserve in Ilo

As planned, on November 15, 2017, Luis successfully completed the round trip from Arequipa to Ilo and back in a single day, as detailed in Chap. 3. To provide further evidence and documentation, he collected samples of guano and captured numerous photographs during his expedition. These photographs can be found in Fig. 9.4, while the map in Fig. 9.3 indicates the locations of Ilo and Arequipa. The samples were delivered to me in Saint-Quentin, France, on November 28, 2017. At that time, Ralph was still in Chile, accompanied by his friend Alain. I immediately opened the box containing the samples and observed that the guano sample had a light brown color.

Fig. 9.4 The Punta Colles
site in Ilo. **a** the entrance to
the reserve; **b** the seabirds on
the rocks; **c** the layer of
guano on the ground from
which Luis took samples
(arrows)

9.5   Analyzing the Guano from Ilo

The examination of the guano sample under our scanning electron microscope on
January 24 and 31, 2018, revealed fascinating insights, as displayed in Fig. 9.5
alongside an optical microscope photograph. The spectrum obtained clearly indicates
the presence of various elements, including carbon (C), nitrogen (N), oxygen (O),
iron (Fe), sodium (Na), magnesium (Mg), aluminum (Al), silicon (Si), phosphorus
(P), sulfur (S), chlorine (Cl), potassium (K), calcium (Ca), and iron (Fe).

Fig. 9.5 Ilo guano; (left) EDS analysis; (right) light microscopy, the scale is 1 mm

To further explore the significance of these findings, we compared the guano analysis with the spectrum obtained from the organic matter found in the Pumapunku andesite, as depicted in Fig. 8.13. I have combined both spectra side by side in Fig. 9.6, revealing that the same elements are present in both analyses, specifically carbon (C), nitrogen (N), oxygen (O), iron (Fe), sodium (Na), magnesium (Mg), aluminum (Al), silicon (Si), phosphorus (P), sulfur (S), chlorine (Cl), potassium (K), calcium (Ca), and iron (Fe).

While the components are indeed similar in both analyses, it is crucial to note that in the monument stone, they are diluted. This is expected, as they are incorporated within the binder used to bind together the grains of andesite sand, as explained in Chap. 7. However, the chemistry remains consistent. The reaction of vinegar (acetic acid) or other carboxylic acids extracted from plants with guano leads to the formation

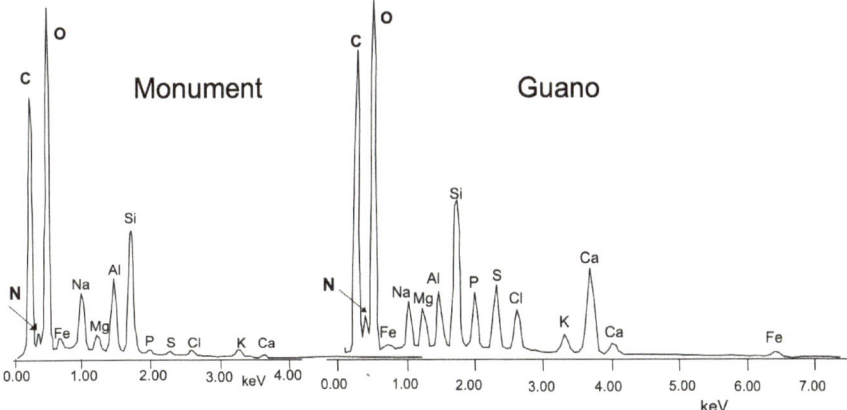

Fig. 9.6 On the left, the EDS spectrum of the organic matter found in the Pumapunku Monument compared with Ilo's guano on the right

of phosphoric and oxalic acids, which are crucial in the production of phosphate-based geopolymers. This chemical process also involves the addition of minerals like alumino-silicates, such as finely altered volcanic tuff, kaolinitic clay, or potentially metakaolin obtained by grinding pottery shards.

It is indeed remarkable that guano, the sought-after hardener, has been identified, completing a journey that began almost 40 years ago with my collaboration with the Peruvian anthropologist Francisco Aliaga. As described in Chap. 1 and mentioned in an article in the New York Times on May 24, 1981, our work has potentially solved one of the most debated archaeological mysteries.

With our newfound understanding, we can now comprehend how the magnificent monuments of Pumapunku, the megalithic terraces of red sandstone, and the remarkable H-shaped structural elements of andesite were crafted using either the techniques of *geopolymers from the mineral world or those derived from the life-giving world*.

While it may seem tempting to conclude this archaeological report at this point, there is still more to uncover. Precisely, I find myself compelled to continue exploring the subject matter. The challenge I posed at the end of Chap. 1 remains in my mind: Is there a connection between the civilizations of the Altiplano, particularly that of Tiwanaku/Pumapunku, and the settlement of Easter Island? Could it be plausible that a civilization originating from the east existed before the Polynesians arrived from the West? Interestingly, this topic is deemed taboo on Easter Island, where expressing such an opinion is discouraged. Nevertheless, I am determined to expound upon and elucidate my hypothesis regarding this matter in the forthcoming chapters.

References

Kolata AL, Ortloff CR (1989) Thermal analysis of the Tiwanaku raised-field systems in the lake Titicaca basin of Bolivia. J Archaeol Sci 16:233–263
Minkes W (2005) Wrap the dead. Archaeological Studies Leiden University 12 Chapters 5.5.2 and 6.5.2, the Netherlands
Towers J (1845) Guano and its analysis. Br Farmer's Mag 9:389–400

Chapter 10
Did the Exiles from Tiwanaku Arrive on Easter Island?

The Vinapu wall made of volcanic rock—Who are these "Exiles from Tiwanaku"?—The places of exile between AD 800 and 900—Exiled in Arica—Why do they go to sea? An order from the sun god?—What was the island like between AD 800 and 900? The millions of Chilean palm trees!

In this chapter, I present a novel hypothesis, drawing upon the wealth of technological, scientific, archaeological, and historical knowledge acquired in Chaps. 2–9 of this book. Through this, I aim to establish a conclusive, well-referenced link between two South American civilizations and Easter Island, suggesting that the initial inhabitants of the island came from the east, predating the arrival of the Polynesians.

Over the years, numerous visitors during the nineteenth and twentieth centuries have been captivated by the striking resemblance between the Vinapu wall construction on Easter Island and the architecture of Cuzco in Peru. Experts, including archaeologists, ethnologists, anthropologists, writers, historians, and filmmakers for cinema, television, and the internet, have engaged in lively discussions on this subject. Some advocate for the notion of an eastern migration, while others maintain that only the Polynesians could have made the first contact with this remote island amidst the vast Pacific Ocean.

The arguments put forth by those who resolutely support the idea of exclusive Polynesian migration are rather curious. Let me quote a few articles published during the twentieth century. For instance, in relation to the connection between the Vinapu wall on Easter Island and the Andean architecture of the Altiplano, ethnologist Handy (1927, p. 329) writes: "(...) *The art of stone building may, of course, have been independently developed in Polynesia... But what probably happened is that during the hundreds of years of very active voyaging, some Polynesians visited America and returned to Polynesia, having seen the Mexican or Peruvian stonework, and possibly bringing a few stone craftsmen with them.*" In other words, it is suggested

J. Davidovits, *Ancient Geopolymers in South America and Easter Island*, SpringerBriefs in Earth Sciences, https://doi.org/10.1007/978-3-031-75336-7_10

that the Polynesians sent teams to engage in industrial espionage among the Incas at an altitude of nearly 4000 m.

Nevertheless, the most persistent preconceived notion is the belief that South Americans lacked the means to construct boats capable of navigating the vast Pacific Ocean. The compelling argument put forth by several authors is that no Amerindian had boats capable of making crossings, such as the voyage to Polynesia. For these authors, the balsa boats of the west coast of South America would soak up water in a few days if they do not come out of the water to dry and would have been quickly submerged and sunk (Dixon 1934, p. 173, Emory 1942, p. 129, Weckler 1943, p. 35, and Buck 1945, p. 11). The prevailing opinion regarding balsa rafts had become an axiom among American ethnologists at that time, influencing the viewpoints of contemporary scholars and casting doubt on the possibility of any eastern arrival, particularly under the influence of the Hawaiians.

However, Thor Heyerdahl's expedition in 1947 had demonstrated that this argument was false. The ethnologist Robert Langton, who was in favor of at least two waves of migration from the east (from Tiahuanaco, for example), wrote in 1997 in the official Easter Island journal, *Rapanui Journal* (Langton 1997):

> (...) In summary, the author's research offers no support whatever for the orthodox view that prehistoric Easter Island was settled solely by Polynesians and in the early centuries of the Christian era. On the contrary, it provides strong support for Thor Heyerdahl's long-held claim that there were two American Indian periods of prehistoric culture and a later Polynesian period (Heyerdahl and Ferdon 1961). In the light of this, the exciting significance of Easter Island in the prehistory of the Pacific Basin as a whole will not be fully realized until the Polynesian-did-it-all theory is well and truly dead and buried.

During my extensive research and writing for Chap. 5, a pivotal moment occurred that has shaped the trajectory of my investigation thus far. It was at this juncture that I came to fully appreciate the immense value of the photographs captured by our team at Tiwanaku, specifically those documenting the base of the Akapena pyramid.

To illustrate the remarkable similarities between the architectures of Tiwanaku and Vinapu, I have included Fig. 10.1, which compares the image displayed in Fig. 5.23 of the aforementioned chapter with the Vinapu wall. It becomes evident that these two structures share notable resemblances despite not being identical. The Tiwanaku construction dates back to around AD 600, during the pyramid's construction period, while the Vinapu wall, as determined by carbon-14 dating, was erected between AD 850 and 950.

It is worth noting that the Tiwanaku architecture is built in red "sandstone," similar to the material employed in constructing the megalithic terraces of Puma-punku. Through our research, we have discovered that it consists of a geopolymer belonging to the mineral world. On the other hand, the Vinapu wall exhibits distinct characteristics that set it apart from its Tiwanaku counterpart.

Fig. 10.1 **a** Retaining wall at the base of the Akapana pyramid in Tiwanaku, AD 600; **b** the wall of the Vinapu wall on Easter Island, AD 850–950

10.1 The Vinapu Wall Made of Volcanic Rock

The composition of the Vinapu wall is primarily volcanic material extracted from the eastern flank of the nearby Rano Kau volcano, as depicted in the map shown in Fig. 10.6. Geologically speaking, this rock is known as "benmoreite" and is classified as lava. During volcanic eruptions, the benmoreite would have erupted and solidified into a fine-grained rock with minimal vesicles or holes. In essence, it shares geochemical and mineralogical similarities with the volcanic rock found in Cerro Khapia, known as andesite, which is the very material used to construct the "H" structures and gates of Pumapunku.

Several geologists have embarked on missions to Easter Island and have published articles and reports exploring the relationship between the archaeological remains and the geological nature of the Rano Kau volcano. For instance, American geologist Patrick McCoy (2014, pp. 5–23), in his article titled "*The dressed stone manufacturing technology of Rapa Nui, a preliminary model based on evidence from the Rano Kau, Maunga Tararaina, and Ko Ori quarries,*" highlights the following: "(...) *The cultural origins of the dressed stone technology of Rapa Nui has been the subject of considerable debate because of the similarities between some of the finest examples on the island [like the Vinapu] and mortarless block masonry in the Andean highlands of South America (…).*"

In their studies, geologists, including McCoy, sought out quarries where the extraction of building blocks could have taken place. However, McCoy encounters a familiar challenge faced by other researchers and similar to the investigations conducted on the Pumapunku/Tiwanaku monuments: they searched for answers in the massive hard rock blocks suitable for construction while the solution lays in the volcanic sand stored at Iwawe. I believe that McCoy and his colleagues on Easter Island may not have found volcanic sand because they did not actively seek it. Nevertheless, based on the nature of the volcanic eruption at Rano Kau, it is plausible to suggest that the phenomenon of "gas-pipes" exists here, just as it does in Cerro Khapia and the *Carbunculus* of ancient Romans. Following the model described in Chap. 7, there would be a juxtaposition of hard benmoreite volcanic rock and volcanic sand, both of the same mineralogical and chemical nature. This would result in a geopolymer with properties akin to the life-giving world, much like the process depicted in Tiwanaku.

Indeed, the presence of volcanic sand, as we have encountered previously at Tiwanaku and in ancient Rome with the *Carbunculus*, holds significant importance. My experiences in studying this matter, along with the insights shared by a group of geologists specialized in volcanic rocks, have provided valuable knowledge in this field.

To delve deeper into the subject, it is crucial for me to establish a precise relationship between the Vinapu and Tiwanaku sites. I firmly believe that there exists a genuine connection between Vinapu and those who journeyed from the east, specifically from South America. Furthermore, my estimation suggests that the people of Tiwanaku arrived at Easter Island between AD 800 and 950. In this chapter, I will meticulously unravel the specific nature of this encounter, shedding light on the intricacies of their connection.

10.2   Who Are These "Exiles from Tiwanaku"?

In Chap. 7, we discussed the transportation of the *piedras cansadas* and the potential time frame in which it may have taken place. I emphasized that during this period, certain events occurred between AD 800 and 900 that are worth considering. Additionally, in Chap. 5, I touched upon the unfortunate mutilation and degradation of one of the andesite gates at Pumapunku, specifically the one adorned with numerous small holes. I explained that the destruction was caused by the gold sheets that once covered the gates, a detail that Cieza de León did not mention in his account. However, while there is a possibility that the conquistadors were responsible for this destruction, archaeologists lean towards a second hypothesis, suggesting that the indigenous people vandalized the site during the general looting of Tiwanaku/

Pumapunku, which occurred around AD 800–900. Interestingly, the Puerta del Sol was spared from destruction precisely because it was not adorned with gold.

I had already mentioned that although we lack written texts to provide definitive proof, anthropologists believe that after AD 800, there was a significant shift in the governance of Tiwanaku. This era saw the rise of a more authoritarian ruling class and the implementation of a policy of social division, contradicting the previous emphasis on duality and social uniformity.

During our discussions, Ralph, Frédéric, and I contemplated the idea that the monuments of Pumapunku/Tiwanaku had already suffered extensive damage long before the arrival of the Spanish conquerors. We proposed the notion that if the temples had been plundered, it is plausible that a portion of the ruling class was executed, while the sacred individuals, such as the priests, were forced into exile. To ensure their permanent removal, a radical solution could have been to set them adrift on a raft made of balsa or reeds somewhere along the Pacific Coast. This would have guaranteed that the new leaders would be free from their influence. I must admit that this idea may seem peculiar and audacious (and somewhat speculative). Still, it aligns well with the historical facts that I have reconstructed and will now further develop.

Let us examine the events and circumstances surrounding the potential exile of these individuals from Tiwanaku.

10.3   The Places of Exile Between AD 800 and 900

In light of the absence of written texts and historical records, our understanding of the Tiwanaku civilization relies heavily on the analysis of ceramics as experimental data. One significant aspect of this analysis is the examination of the evolution of pottery.

10.3.1   The Kero Pottery of Tiwanaku

In Chap. 4, we explored how the Tiwanaku civilization was characterized by its advanced ceramic craftsmanship. I would like to recall the words of anthropologist J.W. Janusek, who stated, "(...) *Tiwanaku 1 dating to AD 500–800, begins with the sudden appearance of a new range of elaborate, decorated red-slipped (redware") pottery. In a sharp break from Late Formative 2 [the civilization before AD 500], virtually everyone now had access to elaborate ceramic vessels for household consumption and feasting (...)*".

This specific pottery is known as "tazon" and, along with the "kero" and the jar, represents the hallmark of the prosperous period of the Tiwanaku civilization before AD 800. See Fig. 10.2 for visual representations of these pottery types.

Fig. 10.2 Tiwanaku pottery; **a** on the left, the kero; **b** in the middle, the tazon; **c** on the right, the jar (Museo Nacional de Arqueología Tiwanaku, La Paz)

Here, the most crucial element for our argument emerges: between AD 800 and 900, these three pottery types appeared in two sites that were far distant from Tiwanaku (refer to Fig. 10.3). The first site is situated in the Cochabamba Valley southeast of Tiwanaku, approximately 320 km away, at an altitude of 2700 m, near the edge of the virgin forest. The second site, which is of particular interest to us, is located westward in Arica, in present-day Chile's Atacama Desert, on the coast of the Pacific Ocean, approximately 400 km from Tiwanaku.

It is important to note that the use of the kero vessel was reserved for the ruling class and priests, who utilized it in various ceremonies where chicha or beer was consumed. The other two pottery types served more everyday purposes, such as a fruit bowl and a water jug. They were not found prior to this period.

Fig. 10.3 The two places of exile in Tiwanaku in 800–900 AD: Arica in the West and Cochabamba in the East

10.3.2 Exiled in Arica

The Chilean archaeologist Mauricio Uribe, affiliated with the Universidad de Chile in Santiago, published an article in 2004 titled "*Tiwanaku ceramics and a jar from the Azapa Valley (Arica, Norte Grande, Chile)*." I found this article particularly valuable as it confirmed that the technologically advanced Tiwanaku ceramics, depicted in Figs. 4.5 and 10.2, were crafted using kaolinite clay. Now, let us uncover the identity of the individuals who arrived in the Azapa Valley, located just 15 km from Arica on the Pacific coast, between AD 800 and 900, carrying Tiwanaku pottery with them. It is likely that they constituted a small group, perhaps consisting of as few as a hundred people.

Within the Azapa cemetery (as shown in Fig. 10.4), around thirty Tiwanaku artifacts were unearthed, including the significant presence of the kero vase, indicating that these objects belonged to prominent Tiwanaku Figures, possibly priests. Uribe (2004) suggests that these individuals were not conquerors or colonizers. In his own words, "*(...) To date, no installations are known which would provide evidence of the introduction of enclaves in a process of conquest or colonization of Arica by peoples from the Altiplano (Tiwanaku) (...).*" Furthermore, he states, "*(...) Although the Tiwanaku presence in Arica... is beyond dispute, this would not imply that the population of the Azapa Valley was integrated into a 'colonial' relationship (...).*" (Uribe 2004, p. 97).

It is evident that they were not traders, soldiers, or conquerors but rather migrants who sought to settle in a place amidst the coastal desert. They had broken off any

Fig. 10.4 Tiwanaku ceramics found in Azapa/Arica; on the top: kero vases; in the middle: "tazon" vase; on the bottom: jars; adapted from Espoueys et al. (1995), Uribe and Agüero (2001)

ties with Tiwanaku and appeared to be exiled priests, accompanied by their spouses, servants, and children. During their arduous journey into exile, they passed through Moquegua (as shown in Fig. 10.3), a satellite site of Tiwanaku, where they may have sought refuge in Temple OMO 10. It is worth noting that this temple was the structure I initially intended to study at the outset of our research. I described our efforts and failures at the end of Chap. 2. I concluded: *"(…) At this point, I think I have enough arguments to dismiss Moquegua and finally mobilize all our energy and research towards Pumapunku, in the Bolivian Altiplano (…)"*.

These individuals did not remain in Moquegua as they were not welcomed there, just like in Ilo. During the period AD 800–900, unrest in Moquegua led to the destruction of the OMO 10 temple due to political turmoil, as discussed in Chap. 2. According to the information provided in M. Uribe's article, the exiled priests from Tiwanaku/Pumapunku purchased pottery in Moquegua and sought a more welcoming place, which they found in Arica (Uribe 2004, p. 88).

10.3.3   Why Do They Go to Sea? An Order from the Sun God?

Now, let us explore why they chose to embark on a journey by sea. One hypothesis revolves around the following understanding: we know that the Temple of Pumapunku, along with the Tiwanaku site, was aligned towards the east, where they celebrated the sunrise. As the sun set, it disappeared behind the mountains. However, in Arica, situated on the Pacific Coast, the situation was different. They had the privilege of witnessing the breathtaking sunset. They must have observed the sun, like a glowing sphere, growing larger and gradually sinking into the ocean, seemingly close to the shore. It is reasonable to assume that these exiled priests, devoted to worshiping the sun, possessed a deep longing to witness the place where the sun bid farewell and to exist in harmony with their creator. For them, this desire likely became more than a mere wish—it transformed into an unwavering obsession. The sun god commanded them to join him. But how could they fulfill this divine order?

Their bodies, accustomed to the living conditions of the Altiplano, possessed an extraordinary energy that may have appeared exceptional to those living at sea level. We have examples in modern times of athletes and sportspeople who engage in physical training in centers located over 2000 m above sea level. With this in mind, we can grasp the physical endurance of these exiles. While a few priests may have sailed on Lake Titicaca near Tiwanaku, venturing on a maritime expedition from there required immense faith in their beliefs.

To convince the fishermen of Arica to take them to the place where the sun appeared to set near the coast, the exiled priests and sun-worshippers of Tiwanaku must have employed persuasive means that surpassed the fishermen's usual coastal voyages on their balsa or reed boats. It is widely acknowledged by ethnologists that the South American coastal fishermen were not known for their exploration or navigation skills, often remaining close to the shore. This notion contradicts the development of my hypothesis. However, it is said that faith has the power to overcome barriers, and it is highly likely that the unwavering faith of the Tiwanaku priests and sun worshippers defied this taboo.

Furthermore, it is worth noting that around AD 1400, during the Inca period, other sun priests, including Kon-Tiki Viracocha, embarked on a journey toward the setting sun in the Pacific to meet their creator. They reportedly discovered two islands, one of which may have been the future Easter Island, also called Rapa Nui by its inhabitants.

As I continued my reasoning, I had to confront this idea, deemed heretical by ethnology and archaeologists, with the reality of the ocean, winds, and currents. Fortunately, Ralph had previously found a program called "Windy," which allows me to track the evolution and direction of winds on the internet through the website. I wondered if it was possible to travel from Arica to Easter Island (Rapa Nui). The answer was affirmative, especially during the months of January to June, which constitute the summer and autumn in the southern hemisphere. As an example, I present a screenshot from Friday, May 22, 2020, in Fig. 10.5.

When departing from Arica, the priests had to initially row westward, potentially using double paddles, in order to move away from the coastal current. Their incredible physical endurance was put to use in propelling the oars or paddles as they aimed to reach the setting sun. After a day or two, they would naturally enter the zone influenced by the winds that would carry them westward, specifically towards Hanga Roa on Easter Island, Rapa Nui, totalizing a distance of 5000 km from Arica. Taking into account the presence of the Humboldt Current and assuming they were not affected by storms, we can estimate that their voyage lasted less than eight weeks. I have used Thor Heyerdahl's contemporary experiment as the basis for my calculations. These will be developed below.

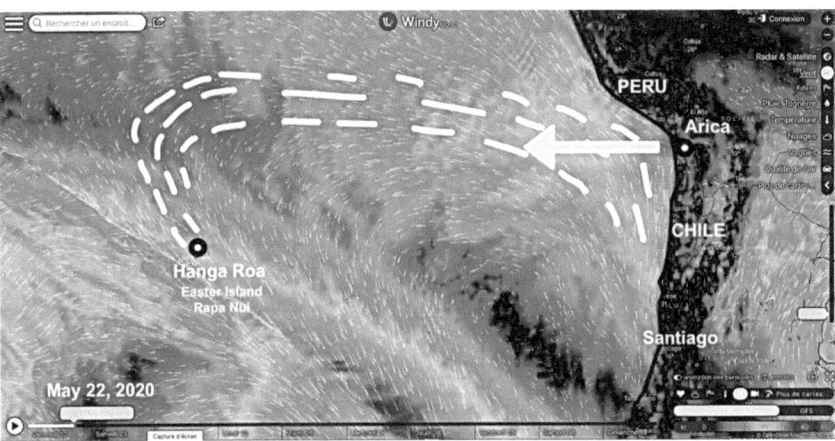

Fig. 10.5 Winds direction on May 22, 2020, in the Pacific Ocean towards Easter Island, Rapa Nui (Hanga Roa). Screenshot from *Windy*. Hanga Roa is the capital of Easter Island, Rapa Nui. The arrow shows the direction of the west, the sunset

Fig. 10.6 Polynesian
V-shaped and
South-American C-shaped
hooks (Easter Island
Museum)

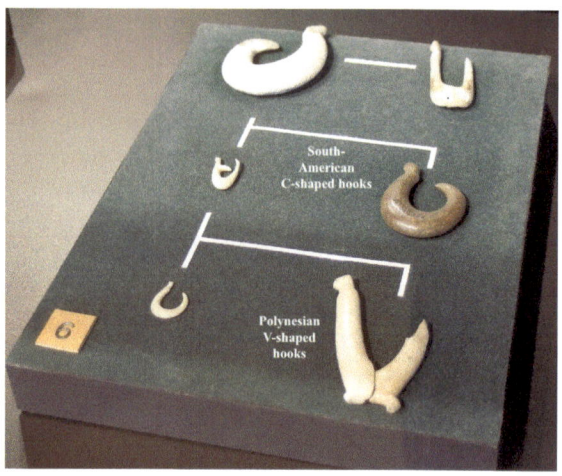

In my research, I was pleased to find literature that supported this route. In Thor Heyerdahl's book, "*Easter Island, The Mysteries Revealed,*" published in 1989, it is mentioned on page 61 that in 1828 and 1834, the Pacific explorer Moerenhout reached Easter Island from Chile. He highly recommended starting the journey from northern Chile, specifically Arica, to take advantage of the southern branch of the Humboldt Current, which would carry ships to the island even in light winds.

It is meaningful to mention the story of the Norwegian ethnologist Thor Heyerdahl (1914–2002) here for those who may not be familiar with it. On April 28, 1947, Heyerdahl, then 33 years old, set sail with a crew of six men on a balsa wood raft from the port of Callao in Peru, aiming to reach Polynesia. The expedition was based on Heyerdahl's theory that the islands of Polynesia were not only inhabited by people from the west but also by people from South America, a theory he had developed during his time on Fatu Hiva Island in the Pacific.

One of the clues that Thor Heyerdahl presented to support his theory was the legend of the Inca Kon-Tiki Viracocha, who sailed westward on a large balsa wood raft, as I mentioned earlier. In the spring of 1946, Heyerdahl presented this theory to a gathering of American anthropologists in New York. However, his idea was met with skepticism, and they challenged him to prove it by sailing from Peru to the Pacific Islands. Undeterred, Heyerdahl believed that the combination of east winds and the Humboldt Current would ultimately guide his balsa raft, named the Kon-Tiki, to Polynesia. After spending 101 days at sea, covering a distance of 4300 nautical miles (7960 km), or an average of 80 km a day, the Kon-Tiki finally landed at the Raroia atoll in Polynesia, proving that people from South America could indeed have reached Polynesia.

Interestingly, more than 75 years later, the idea of such a voyage, particularly one originating from the east, is still considered a taboo topic. However, in the nineteenth century, the route taken by the great sailing ships from Europe to Polynesia was actually from the coasts of South America, sailing from east to west. Over time, though, the prevailing thought shifted in the opposite direction. Nevertheless, this configuration of winds and currents is not uncommon, and with the advent of the internet, we now have the tools to study it in detail. It occurs regularly and leads directly to Easter Island, provided one starts from the right place: Arica, which Thor Heyerdahl did not do. Instead, he departed from further north, from Lima, and his expedition to Polynesia nearly ended in failure. Nonetheless, he held steadfast faith in his hypothesis.

10.3.4 Arica Hooks Found on Easter Island

Among the assortment of tools discovered on Easter Island, there are several types of fishing hooks, including the iconic Polynesian V-shaped hooks (see Fig. 10.6). These hooks, comprising around 70% of the excavated collection, are crafted from a combination of wood and bone (either human or marine mammal), connected by rope. However, it is the stone hooks that often captivate tourists the most. These meticulously carved C-shaped hooks, made from basalt, are often stylized and worn as pendants.

According to Thor Heyerdahl's research (1989, p. 172), there have been reports of stone hooks found on the shores of Lake Titicaca, and ancient Aymara traditions in the Tiahuanaco region indicate their use of pumice hooks for fishing. There are also rarer specimens made from shells, which also take on a C-shaped form. These hooks, identical to those unearthed in prehistoric archaeological sites in the Arica region (Carter 2016; Bird, 1943; Arana 2014), could potentially have been brought by the exiled priests or, more likely, by the Arica fishermen who accompanied them. It is plausible that their descendants on the island continued to employ this traditional South American fishing technique for several centuries.

10.4 What Was the Island like Between AD 800 and 900? the Millions of Chilean Palm Trees!

On September 22, 2016, as I was preparing Ralph's expedition program to Easter Island, Rapa Nui, I came across a fascinating scientific article published in 2003 by Hans-Rudolf Bork and Andreas Mieth, two German scientists from the University of Kiel. The article, titled *"The Key Role of Jubaea Palm trees in the History of Rapanui: A Provocative Interpretation,"* caught my attention. Bork and Mieth (2003) put forth a bold claim that the island was originally abundant with millions of Jubea palms, commonly known as the "Chilean palm," due to its origin in Chile. According to their research, there were an astonishing 6 million palm trees on the island.

What made their idea particularly intriguing was their assertion that the palm trees on Rapa Nui would have contained copious amounts of sap in their trunks, much like the *Jubaea chilensis* species found in central Chile today. This sap is rich in sugar and other vital nutrients. It seems highly unlikely that the Rapanui people would have overlooked such a valuable natural resource. The sap of the Jubaea palm could have been a significant source of sustenance for the island's inhabitants, serving as a crucial liquid in their diet. Bork and Mieth (2003) suggest that not only would a substantial portion of the sap have been used for nutrition, but the surplus of this resource could have been utilized in various other ways as well. Their hypothesis opens the door to intriguing possibilities regarding the multifaceted uses of this valuable palm tree resource.

Upon their arrival on the island, which was not yet known as Easter Island, the priests of the sun and fishermen from Arica were greeted by an abundance of lush vegetation, providing them with palm juice and palm nuts. They wasted no time and began clearing a small area of land. Fascinating pollen analysis studies published in 1992–1993 and mentioned by H-R Bork and A. Mieth shed light on the deforestation process, starting near the Rano Kau volcano in the south, precisely at the location of the Vinapu wall (see the location in Fig. 10.7). Carbon-14 analysis, calibrated to around AD 800–900, provides valuable insights into the Vinapu wall, making it the oldest archaeological remains on the island, predating the arrival of the Polynesians in the twelfth–thirteenth century. According to Bork and Mieth, the deforestation continued from West to East, concluding approximately in AD 1550 at Poike. As I looked into the available data, my research hypothesis continued to gain validation, and new discoveries were made.

It is worth noting that the priests, being the custodians of spiritual, material, and technological knowledge, possessed the expertise to transfer the architectural techniques observed in Tiwanaku to the construction of the Vinapu wall and others

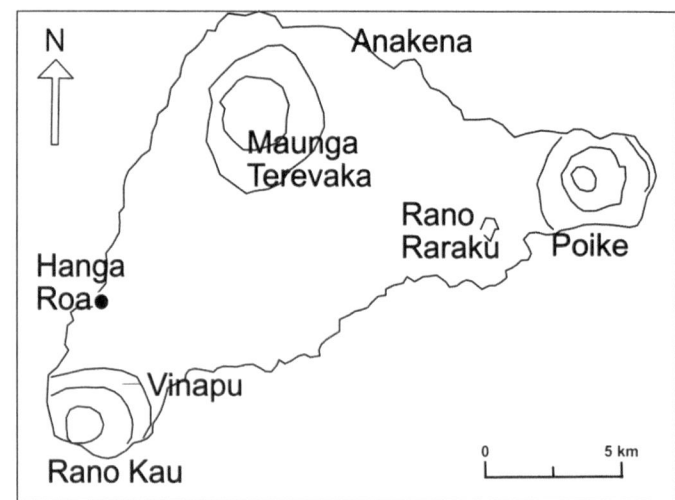

Fig. 10.7 Map of Easter Island, called Rapa Nui, and the main geographical locations, adapted from McCoy (2014)

on the island. The exiled priests had access to an unlimited supply of palm sap, which served as a crucial resource for obtaining the organic acids necessary for creating the geopolymer of the life-giving world. This palm juice is akin to the cactus juice utilized by the indigenous people of Casa Grande in Arizona, as I described in Chap. 9.

In the case of palm juice, the sugar it contains naturally transforms into vinegar (acetic acid) and lactic acid through the action of local bacteria. We will encounter these fossilized bacteria in the analysis of the Ahu Tongariki statues in Chap. 12, where they contribute to acetic and lactic fermentation. I recall the message I sent to Ralph during his visit to the island on November 15, 2016, stating, *"The acid tests are progressing well. After 20 days with vinegar alone, it solidifies. After 3 days, the mixtures of vinegar + lactic acid + citric acid begin to harden. These mixtures, still covered (without drying), act as geopolymer binders."* The addition of guano would further refine the chemical formula. Thankfully, Easter Island played a crucial role in my discovery of this vital reactive element, as discussed in the previous chapter.

The addition of guano, a plentiful source of reactive elements, played a crucial role in perfecting the chemical formula for the Tiwanaku exiles. With this ingredient, they were able to embark on the construction of their temple, using geopolymer blocks made from the life-giving world. It is important to note that the Vinapu wall represents only a fraction of the extensive monuments they built. Unfortunately, most of these ancient blocks were dismantled and repurposed in modern times for the construction of the Hanga Roa harbor docks.

Given that the entire island was once covered in a dense forest of Chilean palms, the search for geological materials like sandstone and kaolin clay was virtually impossible. Additionally, the absence of quality clay meant that pottery-making was not within the Easter Islanders' repertoire. They also lacked natron, a crucial chemical constituent. However, despite these limitations, they possessed a diverse range of resources that allowed them to synthesize geopolymer blocks from the life-giving world.

While the construction of the walls at Vinapu is now attributed to the Tiwanaku exiles, the origins of the hundreds of statues on Easter Island, known as Moai, remain a subject of ongoing debate and speculation. The precise details of who erected these impressive sculptures and the methods they employed are still shrouded in mystery. Various theories propose different possibilities, ranging from the work of the Rapa Nui people themselves to the involvement of outside influences. As I continue my research, I am committed to unraveling the enigma surrounding the Moai and shedding light on their creators.

References

Arana PM (2014) Ancient fishing activities developed in Easter Island. Lat Am J Aquat Res 42(4 Oct 2014):673–689

Bird J (1943) Excavations in Northern Chile. Anthropological Papers of the American Museum of Natural History 38 part 4, New York

Bork HR, Mieth A (2003) The key role of jubaea palm trees in the history of Rapanui: a provocative interpretation. Rapanui J 17(2):117

Buck PH (1945) An introduction to polynesian anthropology. Berice P. Bishop Museum Bulletin, Honolul, p 187

Carter CP (2016) The economy of prehistoric coast of Northern Chile: case study: Caleta Vitor. Ph.D Thesis, The Australian National University, Canberra, Australia

Dixon RB (1934) The long voyages of the Polynesians. Proc Amer Philos Soc 74(3):167–175

Emory KP (1942) Oceanian influence on American Indian culture; Nordenskiold's view. J Polynes Soc 51(2):126–135

Espoueys O, Uribe M, Roman A, Deza A (1995) Nuevos fechados por termoluminiscencia para la cerámica del período Medio en el valle de Azapa (Primera Parte). Actas del XIII Congreso Nacional de Arqueología Chilena, T. II: 31–53. Antofagasta

Handy ESC (1927) Polynesian Religion. Bernice P. Bishop Museum Bulletin, Honolulu, p 34

Heyerdahl T (1989) Île de Pâques, Les mystères dévoilés, Albin Michel, Paris, ISBN 2-226-03529-X

Heyerdahl T, Ferdon EN (1961) Reports of the Norvegian archaeological expedition to Easter Island and the East Pacific. Volume 1, Archaeology of Easter Island, Monograps of the School of American Research and the Museum of New Mexico No. 24, Part 1

Langton R (1997) Evidence for three prehistoric migrations to easter island. Rapanui J 11(1):21–23

McCoy P (2014) The dressed stone manufacturing technology of Rapa Nui: a preliminary model based on evidence from the Rano Kau, Maunga Tararaina, and Ko Ori quarries. Rapa Nui J 28(2):5–23

Uribe M (2004) Tiwanaku ceramics and a jar from The Azapa Valley. Estudios Atacameños, Nr. 27. Universidad de Chile, Santiago, pp 77–99

Uribe M, Agüero C (2001) Alfarería, textiles y la integración del Norte Grande de Chile a Tiwanaku. Boletín de Arqueología PUCP 5:397–426

Weckler JE (1943) Polynesian explorers of the Pacific, vol 6. Smithsonian Institution War Background Studies, Washington, D.C.

Windy, www.windy.com

Chapter 11
Who Built the Enigmatic Statues of the Rano Raraku Volcano?

The enigmatic statues of the Rano Raraku volcano—The Mapuche people and the *Chemamülles* of Chile—Could the statues of the volcano be *Chemamülles* made of stone?—The signature of chemical pollution—The statues of the volcano are indeed *Chemamülles* made of stone.

Clearly, the builders of Tiwanaku/Pumapunku, on the opposite side of the Pacific from Easter Island, were not known for creating stone statues in large quantities. Archaeologists have discovered a few examples, such as the sandstone Fraile statue depicted in Chap. 3 (Fig. 3.4) or the Ponce statue shown in Chap. (Fig. 5.20). However, these statues differ significantly from the various statues found on Easter Island. It is, therefore, reasonable to question who and when these colossal statues were erected. It is worth recalling that the early explorers of the eighteenth century, Jakob Roggeveen and James Cook, believed that the statues were made of artificial stone (as discussed in Chap. 1).

11.1 The Enigmatic Statues of the Rano Raraku Volcano

Let us focus on the statues over 10 m in height, which are embedded into the ground on the slopes of the Rano Raraku volcano, facing the ocean (see Fig. 11.1). Despite assertions from those who support the Polynesian theory, this type of colossus is not part of the traditional customs originating from the western Polynesian islands. Planting statues on the volcano slopes does not align with the traditional Polynesian cult, which involves constructing stone sanctuaries on terraced platforms known as "Ahus." This funerary monument style is well-documented across the Polynesian islands. However, here on Easter Island, we encounter something distinct, leading me to question if this custom may have originated from the east. Nevertheless, we can establish that this practice does not derive from Tiwanaku. Could we be witnessing a second wave of migration from another civilization in South America? Now, I will share the limited information we have at our disposal.

The postcard image in Fig. 11.1a captures Ralph standing next to one of these statues in the traditional pose. In his 1965 book *"Fantastique Ile de Pâques"* French explorer Francis Mazière noted that these statues, with their feet cut to be embedded in the ground, were distinct from the truncated statues that once stood on the large stone platforms. He highlighted the striking differences between them and the statues found on the Ahus. Similarly, during his visit in 1870, Pierre Loti wrote about the two types of statues: the overturned and broken ones on the beaches (see the Moai statues in Chap. 12) and the other, more sinister ones from a different era, still standing in solitude on the other side of the island, those described in this chapter.

Unfortunately, I lack scientific data specifically concerning these statues, unlike the ones found on the Ahus. In her 1973 report to UNESCO, scientist G. Hyvert did not analyze them, possibly due to their intact state and the desire not to cause any damage, unlike the broken ones at Ahu Tongariki.

In 1956, Thor Heyerdahl undertook the task of clearing the perimeter of one statue, as indicated in Fig. 11.1. However, the work did not involve collecting samples from it. Prior to this, it was believed that these statues lacked bodies and only represented busts. However, the cleaning efforts revealed a 10-m-tall body, where the buried portion had a lighter color, as depicted in the image of Fig. 11.1. From an artistic standpoint, these statues are considered the most perfect. They lack eyes and exhibit a slender, elongated structure, giving them a less massive appearance compared to the shorter and rounder statues found on the Ahus. I believe that the existence of these two distinct styles and manifestations is an underestimated aspect in the field of Archaeology.

In 2013, anthropologist Mara A. Mulrooney published an article that presented the results of 313 carbon-14 analyses conducted across various areas of the island (Mulrooney 2013). Surprisingly, none of these analyses pertained to the enigmatic statues scattered on the slopes of the volcano. This left me wondering whether these colossal structures could be contemporary with the Vinapu wall built by the Tiwanaku exiles. Many writers and explorers concur that these statues are the oldest on the island and were created long before the arrival of the Polynesians with their Ahus in the twelfth century. Could they be the first statues constructed by the descendants of the Tiwanaku exiles? Yes and no. I will endeavor to explain why.

One newfound knowledge stems from an unforeseen encounter experienced by Ralph. Upon his return from the Tiwanaku expedition and following a week at his friend's house in Chile, he made a stop in Santiago to visit the Museo Chileno de Arte Precolumbino. On December 6, 2017, he sent me the following message: "*I visited the museum and took numerous pictures that I am sharing with you. You must read the text about the Chemamülles, the funerary statues of the Mapuche people. There was a German guide at the time, and one of the tourists spontaneously remarked in German: <but they look like the statues on Easter Island>* ."

Fig. 11.1 The statues on the slopes of the Rano Raraku volcano; **a** Ralph in front of two statues; **b** the statue excavated by Thor Heyerdahl in 1956 and 1987 (courtesy Thor Heyerdahl 1994); **c** the arrow shows the location of the statue which was excavated and then later re-covered (a small depression in the ground remains)

Fig. 11.1 (continued)

11.2   The Mapuche People and the Chemamülles of Chile

Let us study the text and the accompanying photos (Fig. 11.2). The description of the Chemamülles begins as follows: *"These impressive wooden statues (the Chemamülles) were placed on the tombs of the ancient Mapuche cemeteries. They embody the spirit (am) of those interred there, embarking on their journey to the afterlife. Chiefs and great warriors ventured eastward to inhabit the volcanoes of the blue land of kalfumapu, while others journeyed westward to partake in bitter potatoes beyond the sea."* (Santiago, Museo Chileno de Arte Precolumbino).

The Mapuche people, who originally occupied territories that encompassed parts of Chile and Argentina south of Santiago (as depicted in Fig. 11.3), are the primary ancestral people of Chile. They are recognized as one of the nine original nations in the country's history.

11.2.1   The Chemamülles, Wooden Statues

These remarkable wooden sculptures can stand as tall as 4 m. The ones representing chiefs were placed on the slopes of volcanoes and bear a striking resemblance to the statues found at the Rano Raraku volcano, in contrast to the smaller wooden Polynesian Moai on display at the Easter Island Museum (Fig. 11.2), which have a radically different style from the stone giants. The *Chemamülles* were likely carved from sturdy trunks of pellín oak or laurel, showcasing the textured and irregular nature of the worked wood. The rugged and austere faces with elongated ears in Fig. 11.2 strongly resemble the statues of Rano Raraku.

Fig. 11.2 a The wooden
statues of female Mapuches;
b Wooden Chemamüle
(Museo Chileno de Arte
Precolumbino, Santiago,
Chile, RD 2016);
c Polynesian wooden statue
from Easter Island, 50 cm
high (Easter Island Museum)

Fig. 11.3 Mapuche territory
in Chile and Argentina, in
grey. The arrow shows the
contacts from Arica

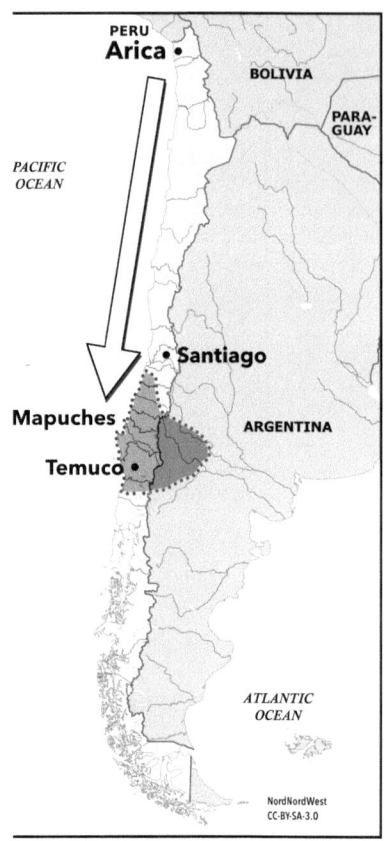

11.2.2   The Tiwanaku Exiles Who Lived Among
the Mapuche?

Tomás Guevara, a Chilean historian, anthropologist, and ethnologist specializing
in the study of the Mapuche people, proposed in 1925 (*Historia de Chile: Chile
prehispano*, Volume 1) that there may have been some historical contact between the
Mapuche people and the northern peoples, specifically those from Arica. Guevara
suggests that various groups of fishermen with ties to Tiwanaku culture migrated
from north to south, greatly influencing the culture of the Mapuche people (see
Fig. 11.3). Should I pursue this path? It is worth noting that Tomás Guevara's thesis
has sparked considerable controversy among Chilean historians.

  However, there are similarities between Mapuche traditions and those of
Tiwanaku. In Chap. 8, I connected the geopolymer of the life-giving world with
the principles of life in Tiwanaku, as described by American anthropologist Janusek
(2008). I wrote: "*(…) The bluish-grey andesite evoked the life-giving principles of
distant sacred volcanic peaks revered by numerous communities in the altiplano*

while symbolizing the complementary life principle of water in Lake Titicaca." The water of Lake Titicaca is also blue, and this color parallels the description of the land where the *Chemamülles* were placed, as mentioned in the notice at the Santiago Museum: "*Chiefs and great warriors went to the east to inhabit the volcanoes of the blue land of kalfumapu (...)*." In both cases, the land of the volcanoes is described as blue.

It seems that we can make an assumption that some of the exiles from Arica were indeed part of this migration towards the south, as depicted in Fig. 11.3.

To further support this hypothesis, I suggest exploring the website Windy, where we can examine the currents and winds to determine if it would have been possible for individuals to reach Easter Island, Rapa Nui (Hanga Roa), with their assistance. In Fig. 11.4, the map reveals that during the spring in the southern hemisphere (November), the winds are highly favorable. They provide a direct route from Temuco, located south of Santiago in the Mapuche region, leading straight to Easter Island (Hanga Roa), 3800 km distant. These winds even make a return journey feasible.

It is interesting to note that sailing from the Chilean coast to Easter Island presents two options depending on the seasons. In the summer and autumn, one would depart from Arica in the north. Conversely, in winter and spring, the journey would commence from Temuco, within the Mapuche territory in the south.

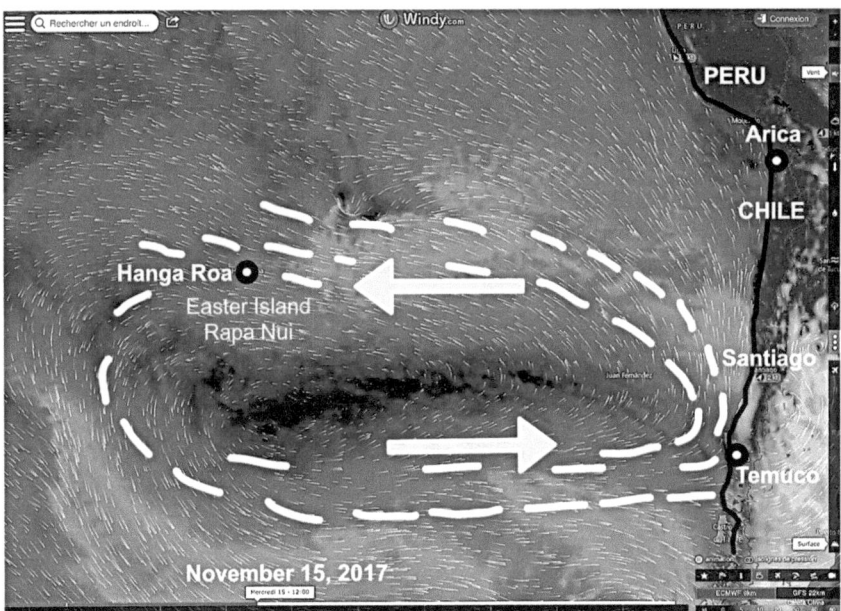

Fig. 11.4 Map of November winds in the Pacific Ocean towards Easter Island, Rapa Nui (Hanga Roa) from the coast of Chile, south of Santiago

In the event that we are unable to confirm the hypothesis involving exiles from Arica arriving from the north with the Mapuche fishermen, there remains a second possibility. We can reasonably consider that groups of Mapuche fishermen might have reached the island, either voluntarily or involuntarily, following a storm or similar circumstances.

11.3   Could the Statues of the Volcano Be *Chemamülles* Made of Stone?

In both scenarios, whether it is the exiles from Arica and their Mapuche companions or the Mapuche themselves, they safely arrive on Easter Island, marking the second wave of people coming from the east. Once there, they encounter the descendants of the initial Tiwanaku exiles from Arica, but this meeting takes place 100–200 years later. In order to uphold their tradition, the Mapuche fishermen must select one of the island's volcanoes as the designated site for their *Chemamülles* dedicated to their chiefs. Ultimately, they decide to choose Rano Raraku. However, they face an insurmountable problem. Palmwood, the only large tree found on the island, is unsuitable for sculpting due to its fibrous nature. There are no oak or laurel trees available, and while there is a shrub called *Sophora toromiro* (toromiro) with a hardwood trunk, its trunk is too short to create *Chemamülles*.

If the Mapuche people who have settled on the island wish to continue their worship of *Chemamülles*, they must find an alternative solution. Could they have crafted *Chemamülles* from artificial stone, using the geopolymer of the life-giving world developed by the Tiwanaku exiles from Arica to construct the Vinapu Wall?

The wooden *Chemamülles* share the same stylistic characteristics as the stone statues from Rano Raraku. A simple comparison between Figs. 11.1 and 11.2 reveals the striking resemblance between them. The statues from this period have a shape reminiscent of a large, cylindrical, and uniform tree trunk, similar to the trunks used to carve the *Chemamülles*. These statues are inserted into the ground and cannot be placed on top of the Ahus.

This was an intriguing hypothesis, and until a few months before, I believed that I had reached the limit of my reasoning. I lacked the crucial information derived from scientific analyses that would allow me to determine the natural or artificial composition of the rock used for the statues. However, I recently came across a study published in December 2019 titled "New excavations in Easter Island's statue quarry: Soil fertility, site formation, and chronology" by Sarah C. Sherwood, Jo Anne Van Tilburg, Casey R. Barrier, Mark Horrocks, Richard K. Dunn, José Miguel Ramírez-Aliaga, in the Journal of Archaeological Science (Sherwood et al. 2019).

The study mentioned in the title focused on the environment surrounding the statues within the Rano Raraku volcano's caldera, including a pond with varying water depths dependent on the climate. Currently, the pond is approximately 2–3 m

Fig. 11.5 a At the top, the caldera or interior of the Rano Raraku volcano, its pond, and marked by a white dot, the location of the statues; the arrow indicates the place where Jo Anne Van Tilburg's team from UCLA carried out their studies; **b** at the bottom, some stone statues inside the volcano

deep, as depicted in Fig. 11.5, along with the positions of a few stone statues indicated by white dots.

The arrow on the figure indicates the specific location where Jo Anne Van Tilburg's team from UCLA (University of California, Los Angeles) conducted their research, uncovering two statues (no. 156 and 157) to a depth of 5.5 m from the ground surface. What caught my attention in this article was the new chronology it presented, which could provide insights into the period when some of these statues were placed on the inner slope of the volcano.

Carbon-14 dating of organic elements found in the soil of the two statues suggests a timeframe around AD 1455. However, activities in other areas of the caldera would have started earlier. It is believed that the construction of the first stone statues, those erected on the outer slopes of the volcano (refer to Fig. 11.1), began much earlier, around AD 1100–1200 (see, for example, in the next chapter at the end of Sect. 12.2).

When the research paper was published in December 2019, the University of California's press release focused on its agricultural implications. It included statements like: "*Rapanui people likely believed the ancient monoliths helped food grow on the Polynesian island, study reveals.*" Initially, I had no interest in this type of information, so I did not pay much attention to the content of the article. I merely read its summary. However, one year later, while writing this chapter, I revisited the

UCLA press release and came across a statement from the first author of the article that intrigued me. Sherwood said, *"When we got the chemistry results back, I did a double-take. There were really high levels of things that I never would have thought would be there, such as calcium and phosphorous."* This piqued my interest, and I made an effort to read and comprehend the 22-page article.

11.4   The Signature of Chemical Pollution

Upon further examination of the article, I realized that the title, *"New excavations in the Easter Island statue quarry: Soil fertility, site formation, and chronology,"* was misleading and served as a distraction from the actual scientific content. As I delved deeper into the text, I came across some astonishing statements, and I have selected five of them to highlight:

1. On page 4, it states, *"(…) The tuff is especially suitable for carving but not for preservation due to its relatively high porosity, low bulk density, and homogeneity across the area."* According to the article, more than 90% of the statues have been "carved" from this volcanic tuff. This raises questions about why the statues appear unaffected by the severe erosion of the volcanic tuff on the cliffs of the volcano. Is it possible that the statues are made from a different type of stone altogether?

2. On page 10, regarding the presence of phosphorus (P) mentioned by Sarah C. Sherwood in the UCLA press release, it states, *"(…) While P is typically the least mobile among soil elements (…) noting the highest P results are derived from 285 to 360 cm in an area of incipient surfaces, then human activity is likely the primary contributor to these higher P levels."* Fig. 11.6 illustrates that the maximum value for phosphorus (P) is 450% higher than the recommended value. To reach such exceptionally high levels of phosphorus (P), it is plausible that residents must have added significant amounts of a product containing this element. This suggests that it is not a natural occurrence.

3. On pages 11–12, it states, *"(…) The phytolith assemblages throughout the profile are dominated by palms (up to > 70%) (…) The large to very large amounts of palm phytoliths throughout the entire profile could seem at odds with the coincident low pollen values for this taxon (…)."* Phytoliths are mineral elements composed of silica that are typically found in the fibers of certain plants, in this case, palm trees. The absence of significant pollen traces indicates that no nearby palm trees are growing naturally. However, the abundance of palm phytoliths suggests that large quantities of palm extracts, such as palm juice and/or palm ash, have been imported from outside sources, carrying their distinctive phytoliths.

4. Furthermore, on page 16, it is mentioned, *"(…) Below 2 m, the profile reveals high exchangeable calcium Ca, thus indicating the inverse to most stable volcanic soils where typical weathering in tropical and subtropical conditions results in cation depletion (…)."* However, Fig. 11.6 shows that the maximum value of

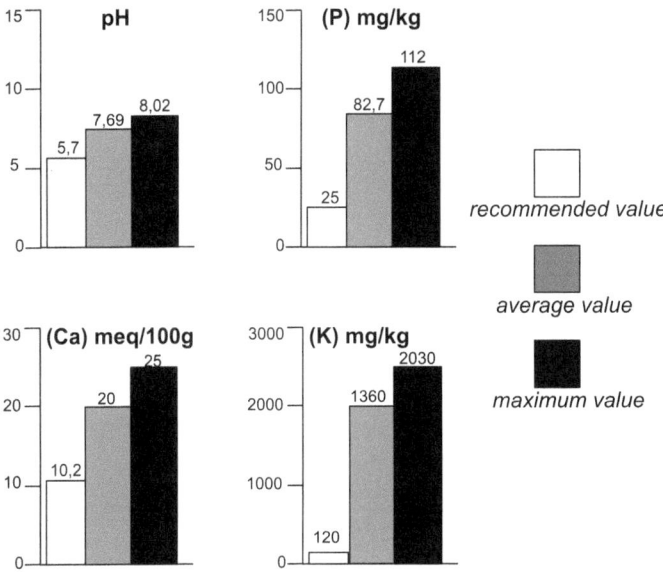

Fig. 11.6 Recommended, average, and maximum values measured for pH, phosphorus (P), exchangeable calcium (Ca), and potassium (K) in the ground, statue Nr. 156, caldera of the Rano Raraku volcano, according to Sherwood et al. (2019)

exchangeable calcium (Ca) is 250% higher than the recommended value. This presence of such large amounts of calcium (Ca) is associated with the phosphorus (P) mentioned in statement No. 2.

5. Another noteworthy statement can be found on pages 16–17, which states, "(…) *Potassium (K) is an especially essential element (…) The Rano Raraku K levels range from 812 to 2,030 mg/kg (m = 1360). The New Guinea study interpreted exchangeable K as low (< 50 mg/kg), medium (< 50–120 mg/kg), and high (> 120 mg/kg), demonstrating the phenomenal levels of available K at Rano Raraku.*" The potassium (K) levels observed are 1000–2000% higher than the typical value of 120 mg usually found in soil (Fig. 11.6). This is an exceptional and abnormal occurrence.

To provide clarity, I have compiled the average and maximum values of these chemical elements in Fig. 11.6. They are presented alongside the recommended values for soil chemicals necessary for successful intensive agriculture in the Polynesian region, as determined by the Hawaii Biocomplexity project and the New Guinea study mentioned in the UCLA paper. It is also worth noting that the results indicate an alkaline pH range of 7.69–8.02 for the soil, whereas, for this type of volcanic tuff, the pH should typically be acidic, around 5.7 (Sherwood et al. 2019, p. 14).

Upon further analysis of the scientific article, it becomes evident that these measured values are not ordinary but rather the result of long-term contributions made by the island's inhabitants. While the authors of the research claim that they aimed to enhance soil fertility and improve agricultural productivity within the crater, I have serious reservations about this assertion.

The excessive levels of chemical elements observed are counterproductive and potentially detrimental to agriculture. In fact, the quantities are comparable to what one might find in cases of industrial soil pollution caused by the use of chemical products, which can have more harmful than beneficial effects on plant growth. The emphasis on agricultural applications in the press release from UCLA is, therefore, erroneous, and it misrepresents the true implications of the scientific findings. This phenomenon of chemical pollution warrants an alternative explanation, which I intend to outline in this chapter.

11.5   The Statues of the Volcano Are Indeed *Chemamülles* Made of Geopolymeric Stone

The question of how the Easter Islanders managed to transport water and produce geopolymer mixtures has always intrigued me. Upon analyzing the data mentioned above, it becomes apparent that the interior of the Rano Raraku volcano served as a workshop for creating the "concrete" using geopolymeric materials from the surrounding environment. The Mapuches, who had settled in the area, replaced their traditional wooden *Chemamülles* with stone versions.

While I have not come across any scientific mineralogical analysis conducted on stone *Chemamülles*, I now possess the necessary information to describe the process of mixing the "concrete" within the volcano. Water was readily available, along with volcanic tuff that had disintegrated. The only missing components were the essential chemical elements. These included organic acids derived from the maceration of palm juice, guano as a hardening agent (containing calcium phosphate and calcium oxalate), palm wood ash (potassium carbonate and phytoliths), as well as other minerals yet to be determined. The resulting geopolymeric stone paste was then transported in baskets to the statue construction site located on the outskirts of the volcano.

This recent scientific paper provides compelling evidence that the Easter Islanders likely employed geopolymer chemistry in the construction of their statues. It sheds light on their innovative methods and adds to our understanding of the remarkable achievements of this ancient civilization.

References

Mulrooney MA (2013) An island-wide assessment of the chronology of settlement and land use on Rapa Nui (Easter Island) based on radiocarbon data. J Archaeol Sci 40(12):4377–4399

Sherwood SC, Van Tilburg JA, Barrier CR, Horrocks M, Dunn RK, Ramírez-Aliaga JM (2019) New excavations in Easter Island's statue quarry: soil fertility, site formation and chronology. J Archaeol Sci 111:104994

Windy, www.windy.com

Chapter 12
Who Built the Moais Statues of Ahu Tongariki?

The 1973 UNESCO analytical report—Fossilised microorganisms and bacteria contained in the statues—Volcanic tuff sand found at the base of the Ahus.

When the Polynesians arrived on the island around AD 1200, they encountered a population that had been creating geopolymer statues representing the life-giving world. The Polynesians initially constructed their platforms, known as Ahus, following their ancestral traditions or by adapting the technique used in the Vinapu wall. Realizing the relative ease of implementing this method, they further developed it to manufacture their own unique style of statues, the Moais, modifying the shape and design. Some of these Moais grew to colossal proportions, showcasing the skills they had acquired to produce and erect them on the Ahus. These new statues were distinct and unlike anything their Polynesian ancestors had carved before.

Let us begin with a classic postcard view (Fig. 12.1). Ralph stands before the Ahu Tongariki, situated on the ocean shore near the plateau and the Poiké volcano (refer to the map in Fig. 10.6). I deliberately chose this Ahu because its 15 Moais hold some of the secrets we need to understand the manufacturing technique. To guide us, I will rely on the only comprehensive report published in 1973 concerning the mineralogical and petrographic composition of Easter Island statues. This report, commissioned by UNESCO, was conducted by Gisèle Hyvert, a French scientist in the Museum of Natural History in Paris and an expert in the restoration and preservation of ancient monuments.

While there are other superficial analyses of small samples taken from various statues, which primarily aim to compare the nature of the crystalline grains that compose them and determine their origin from the Rano Raraku volcano, these studies are of limited use.

For instance, I would like to mention the study by Van Tilburg et al. (2008) titled "*Petrographic analysis of thin-sections of samples from two monolithic statues (Moai), Rapa Nui (Easter Island),*" which states, "*(…) Both samples displayed textures and mineralogy characteristic of the Rano Raraku tuff deposit, and are*

Fig. 12.1   Ralph in front of the Ahu Tongariki and its 15 Moais

characterized as an unwelded sideromelane tuff with moderately low alteration to palagonite (...)." However, these authors only examined the structure of sideromelane, a volcanic glass present in the palagonite tuff, which is a tuff containing clays and zeolites that they neglected to analyze. This observation should be compared with the discussion in Chap. 8, regarding Fig. 8.7, where I concluded: "*(...) In short, this was the definition given by someone who did not know the terminology of what I refer to as a geopolymer (...).*"

12.1   The 1973 UNESCO Analytical Report

Indeed, I recall mentioning this report in Chap. 1, as it holds valuable information. In May 1974, I reached out to UNESCO in my pursuit of precise information. Following an interview with the head of the UNESCO Action Service for the Safeguarding of World Monuments and Sites, I obtained the report of the Expedition organized by UNESCO in 1972 to Easter Island. This report includes mineralogical analyses of the statues, which could potentially provide scientific evidence supporting the notion of their manufacture through agglomeration. Take a look at Fig. 12.2 for reference.

The rocks comprising the Moai statues exhibit two types of alterations: mechanical and chemical. Hyvert (1973) addresses this matter, stating, "*(...) chemical alteration which results in the formation of a white deposit that is primarily composed of opal. This indicates the remobilization of silica from the volcanic glass. Diatoms have been observed in these silica layers, although it should be noted that these organisms are fossilized, with no living diatoms found in the examined samples. (...) The presence of chlorine in the clay cement, identified through scanning electron microscopy, suggests a marine influence during the chemical weathering processes that affected the rock (...).*"

Fig. 12.2 a The cover of the
UNESCO report by Gisèle
Hyvert; **b** SEM photos of
barrel-shaped fossilized
microorganisms (statues of
the Ahu Tongariki), 5-µm
scale, from Hyvert (1973)

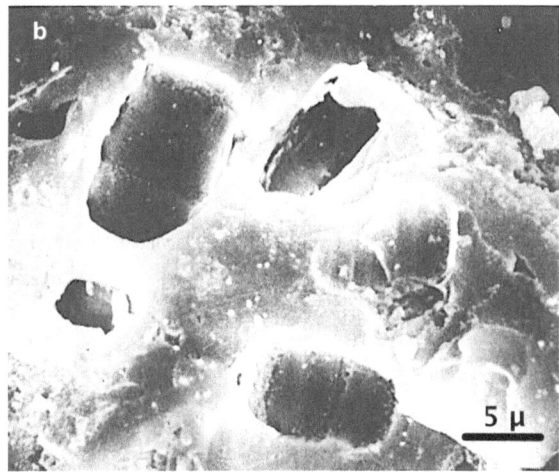

This statement offers two significant pieces of information:

- White silica deposits are formed, within which fossilized microorganisms, as depicted in the photo shown in Fig. 12.2, are discovered.
- The presence of chlorine (Cl) raises exciting considerations. Given our proximity to the sea, one might initially attribute the chlorine to the influence of sodium chloride (NaCl) present in seawater, potentially resulting from the action of sea

sprays on the stone. However, astute readers who have followed my argumentation thus far will recollect that in both Chaps. 8 and 9, the same chlorine (Cl) was identified as part of the organic matter contained in the andesite-made artificial stone, as well as in the guano, as depicted in Figs. 8.13 (Chap. 8) and 9.6 (Chap. 9). Therefore, chlorine may also be regarded as a characteristic sign of the artificial nature of the rock.

The circumstances under which G. Hyvert conducted her mission were truly extraordinary. Due to past events, all the statues had been toppled from their positions on the Ahu. Additionally, during the devastating tidal wave of 1960 that ravaged the bays of Easter Island, the statues were further dislodged and overturned, resulting in them breaking into multiple pieces. This unfortunate event, however, presented an opportunity for Hyvert to gather numerous samples for analysis. Initially, the site housed a total of 20 statues, but only 15 could be restored and placed back on Ahu Tongariki starting in 1990. This restoration project was made possible through the generous funding provided by the Japanese company Tadano.

12.2   Fossilised Microorganisms and Bacteria Contained in the Statues

According to G. Hyvert's observations, it is worth noting that the majority of the statues on Easter Island display extensive damage and fragmentation. These statues also exhibit a significant buildup of white deposits, forming a crust that adheres to the rock. At the base of the opal layer, near or in contact with the rock, there is a prominent abundance of small rectangular diamond-shaped structures resembling barrels, as depicted in Fig. 12.2.

These structures are believed to be either microorganisms or remnants of bacteria colonies, which could potentially be linked to the production of geopolymer stone through two possible scenarios.

In the first scenario, these microorganisms may have played a role in the biological fermentation process that converts sugar in palm juice into organic acids such as acetic acid (vinegar) and lactic acid. These acids could have been utilized in the production of geopolymer of the living world. The second scenario suggests that these bacteria could have evolved as a result of chemical pollution in the soil, as discussed previously.

This research reminded me of another study published in 1998, which has been sitting on my shelves for quite some time. The study involved a team of European scientists from Belgium, France, the UK, and the Netherlands, led by Henri J. Dumont, published in the *Journal of Paleolimnology* (Dumont et al. 1998). Their investigation focused on coring at a depth of 4 m in the sediments that were washed into the lake of the Rano Raraku volcano during specific periods in the past, for example, in the fourteenth century AD. At a depth ranging from 3 to 3.10 m, they discovered a significant amount of pollen from the native *Chilean Jubea* palm tree but very little from other species like grasses. Interestingly, this palm pollen decreased significantly until a second level between 2 and 2.4 m and then disappeared entirely above that level. However, at the same core level of 3 m, the study also identified the presence of *Cyanobacteria*, indicated by the signature pigment of zeaxanthin and echinenone.

It is known that *Cyanobacteria* thrive in environments polluted by human activity, resulting in eutrophication or dystrophication. These bacteria often manifest as colorful blue-green algal blooms in bodies of water. The presence of excess nitrogen (N) or phosphorus (P) in the water, often caused by intensive farming or inadequate wastewater treatment in urban areas, can contribute to the growth of these algal blooms. In the case of Easter Island, the Sherwood study (Sherwood et al. 2019) provided evidence of an excess of phosphorus (P; see the details in the previous Chap. 11). Additionally, nitrogen (N) pollution can be attributed to the extensive use of guano as a hardener, which contains high amounts of ammonium salts (oxalate and phosphate, as indicated in Table 9.2 in Chap. 9).

Accordingly, these bacteria were intricately present in the water used during the manufacturing process of the geopolymer paste. This discovery suggests that the statues found at Ahu Tongariki, much like the stony *Chemamüilles*, may indeed be crafted from an artificial form of rock. The living *Cyanobacteria*, with their barrel-like appearance, closely resemble the fossilized bodies depicted in Fig. 12.2.

Furthermore, G. Hyvert adds observation regarding the presence of calcium oxalate, which is explicitly related to whevellite. This finding suggests the past presence of lichens or the potential existence of fungi. However, it is essential to note that the biological alteration is minimal, and lichens are unable to establish themselves on this fragile rock. Only a few rare cases of lichen presence have been observed.

Therefore, the calcium oxalate cannot be attributed to lichens since their presence is practically non-existent. Instead, it could potentially be derived from the guano substances, which also include calcium oxalate and aid in the solidification of the malleable stone, as discussed in Chap. 9. It is crucial to highlight that these analyses were conducted on the interior of the rock, eliminating any risk of surface contamination. This same argument was employed in examining the organic matter found in the "andesite rock" of Pumapunku, as discussed in Chap. 8 (Fig. 8.13). It was this very argument that convinced the editors of the scientific journal *Ceramics International* to publish the article announcing our discovery, as mentioned at the end of Chap. 8.

In G. Hyvert's investigations, no living microorganisms responsible for the biological alteration of the rock were encountered. When her analyses were conducted in 1972, these bacteria were already fossilized. Since they do not naturally occur in volcanic rock, it can be inferred that they were introduced into the geopolymer cement during the manufacturing process of the statues, which took place 600–800 years ago. Additionally, the report reveals other significant findings, such as the presence of calcium phospho-carbonate, which is also found in guano. These results further validate the existence of an artificial stone created through a geopolymer system closely tied to the world of life.

Unfortunately, carbon (^{14}C) dating could not be performed for sediment cores deeper than 1.35 m, as the later intervals were assigned to the AD 1300–1450 era.

At a deeper level, the data becomes unusable for establishing a reliable chronology due to the occurrence of a rather massive inwash of carbon of infinite age as part of quarry debris. It is known that even a small amount of modern carbon can make a sample appear older than its actual age by a magnitude of 10,000 years or more.

Nevertheless, the authors of the study suggested an alternative theoretical sediment accumulation rate of 0.85–1 cm per year. The 165 cm thick sediment found between the levels of 1.35 m and 3 m would give an age interval of approximately 195–165 years, tentatively dating back to the interval of AD 1105–1135 to AD 1255–1285.

12.3   Volcanic Tuff Sand Found at the Base of the Ahus

The notion that the statues were constructed directly on the Ahu gains support from James Cook's observations, as mentioned in Chap. 1 (Cook 1777). Cook marveled at how the islanders, devoid of mechanical power, could raise such colossal figures and position the large cylindrical stones on top of them. However, if the stones were indeed artificial, it is plausible that the statues were assembled in their current location, as Cook had proposed.

If this hypothesis holds, one would expect to find remnants of materials such as volcanic tuff sand, soft stone paste from the caldera, and fragments of tuff at the base of the Ahus. Belgian archaeologist Nicolas Cauwe has been advocating for this idea since 2011, aiming to dispel the mysteries surrounding the transportation of the statues by employing logical reasoning (Cauwe 2011 and 2018). Cauwe suggests that the Pascuans dragged roughly carved stone blocks and completed the statues directly on the Ahu. In 2013, he wrote: "*In front of these monuments, there is always a layer of tuff dust (the material from which the statues are made), which tends to prove that a minimum of finishing was done after transportation. Until the precise nature of the transported objects is determined, it is futile to attempt to recreate any transportation techniques. Moving a finished and delicately detailed statue is not the same as transporting a partially sculpted block, which is much less fragile*" (Cauwe et al. 2013, p. 28).

In addition to his written work, Cauwe delivers lectures on his findings. In a recent video entitled Mythes et Mystères de l'Ile de Pâques, (in English: Myths and Mysteries of Easter Island) he explains that the Pascuans would extract uncarved blocks from the quarry and shape them at the intended erection site (Cauwe 2018). Cauwe also discovered substantial quantities of tuff dust underground at the construction level, providing further evidence to support my interpretation.

The volcanic "tuff dust" is akin to the andesite sand found in Stratum V at the port of Iwawe on the shores of Lake Titicaca near Tiwanaku. In Chap. 7 (Fig. 7. 3), I referenced the work of North American anthropologist William Isbell, who concluded that this layer was composed of andesite volcanic sand resulting from the intensive sculpting of blocks sourced from the Cerro Khapia volcano.

Indeed, both N. Cauwe's research on the Easter Island statues and W. Isbell's investigations into the andesite "H" shaped structures and Pumapunku gates follow a similar line of reasoning. According to their findings, the blocks used in these constructions were meticulously cut and carved from the natural volcanic rock, resulting in the sand or "dust" that is found at the sites. However, it is essential to note that this "waste" material does not include the large chips or debris one might expect from the stone-cutting process and, therefore, cannot be considered actual waste.

In the context of our artificial rock theory, this so-called "waste" material actually serves as the raw material collected from the volcanic source. Just as at Tiwanaku, the volcanic tuff sand on Easter Island would have been transported in baskets from the Rano Raraku volcano and stored at the base of the Ahus, where N. Cauwe made his discoveries. The Moai statues were then crafted on the Ahus site, utilizing the ingredients of the geopolymer system derived from the life-giving world.

It is intriguing to observe that in both cases, Pumapunku/Tiwanaku and Easter Island, volcanic materials (rocks and volcaniclastic sediments) are involved, and the identification of biological organic matter undeniably supports the theory of geopolymeric artificial manufacturing. This same technique was employed in the construction of the Moai statues, the Vinapu wall, and the colossal *Chemamülles* statues planted on the slopes of the Rano Raraku volcano.

References

Cauwe N (2011 and 2018) Île de Pâques, Dix années de fouilles reconstruisent son histoire. *Le Grand Tabou*, Versant Sud, Bruxelles, p 34

Cauwe N, De Dapper M, Coupé D (2013) Suicide écologique à l'Île de Pâques: ce qu'en dit l'archéologie. Science et Pseudo Sciences 305(juillet 2013):18–42

Cauwe N (2018) Mythes et Mystères de l'Ile de Pâques. Available at: https://www.youtube.com/watch?v=eyImESm2VzE . Accessed 12 Aug 2024

Dumont HJ, Cocquyt C, Fontugne M, Arnold M, Reyss JL, Bloemendal J, Oldfield F, Steenbergen Cees LM, Korthals HJ, Zeeb A (1998) The end of moai quarrying and its effect on Lake Rano Raraku, Easter Island. J Paleolimnol 20:409–422

Hyvert G (1973) Îles de Pâques, Les statues de Rapa Nui, Conservation et restauration. Rapport UNESCO, 2868/RMO.RD/CLP, Paris

Sherwood SC, Van Tilburg JA, Barrier CR, Horrocks M, Dunn RK, Ramírez-Aliaga JM (2019) New excavations in Easter Island's statue quarry: soil fertility, site formation and chronology. J Archaeol Sci 111:104994

Van Tilburg JA, Kaepler AL, Weisler MI, Cristino C, Spitzer A (2008) Petrographic analysis of thin-sections of samples from two monolithic statues (MOAI), Rapa Nui (Easter Island). J Polyn Soc (NZ) 117(3):297–300

Chapter 13
They Came from America to Build Easter Island

"And in this new world of the Indies, where no letters have been found, we remain ignorant about many things." These were the words of Cieza de León, a prominent Spanish chronicler, in 1553, as expressed in his Chronicle of Peru, which I discussed in Chap. 4 under the title: "Let us try to understand the history of Tiwanaku/ Pumapunku."

Both civilizations examined in this book, the South American civilization in Tiahuanaco, Bolivia, in the Altiplano, which flourished between AD 600 and 800, and the Easter Island civilization, which emerged from AD 800 to 900, lacked a writing system to record their history and pass it down to future generations. Complete silence prevailed, and I had to rely on ingenuity to "become aware of many things." Through this process, I managed to connect the seemingly "impossible" and the "forbidden." The impossible was demonstrated by establishing a solid connection between the Andes of South America and Easter Island, while the forbidden involved establishing that the first inhabitants of the island were Americans, predating the arrival of the Polynesians. It was the Americans who came to build Easter Island.

Throughout my research, I have consistently adopted a comprehensive approach, integrating knowledge from Anthropology, Archaeology, Geology, Mineralogy, and Chemistry. It is my belief that seeking satisfactory answers solely within one discipline would not allow us to comprehend and explain the remarkable journey of *Homo sapiens*, forced to thrive in extreme environmental conditions. How is it possible that after the collapse of the magnificent Tiwanaku civilization, which thrived harmoniously at an altitude of 4000 m in the Andes of South America, a small group of exiles managed to replicate their extraordinary technological expertise on a remote Pacific island like Easter Island? How were they able to construct and erect those colossal statues, such as the *Chemamülles* and the *Moais*, created from artificial stone?

Nevertheless, in order to understand the history of the Altiplano and the surrounding civilizations near Lake Titicaca, anthropologists soon realized that studying the evolution of pottery, ceramics, and clay work was essential. I must acknowledge the contributions of influential anthropologists in this field, starting with John Wayne Janusek from Vanderbilt University in the USA and his book *Ancient Tiwanaku*, published in 2008. Through his work, I discovered the abrupt transition around AD 500, shifting from the production of simple terracotta to the creation of what I described as "high-tech" ceramics in Chap. 4. Janusek (2008, p. 22) wrote in his book, "*(…) Tiwanaku 1, dating from AD 500–800, begins with the sudden appearance of a new array of intricate, decorated red-slipped pottery. There was a clear departure from the Late Formative 2 period [the civilization prior to AD 500], with almost everyone gaining access to elaborate ceramic vessels for everyday use and feasting (…).*"

This high-tech ceramic from Tiwanaku became a crucial tool for precise dating throughout various chapters of this book. In Chap. 7, thanks to the research conducted by another anthropologist, William Isbell, from the State University of New York at Binghamton (USA), this same "high-tech" ceramic enabled us to identify and understand the volcanic material employed in the andesite geopolymer rock, which forms the imposing gates and "H" structures of Pumapunku.

In Chap. 10, thanks to the diligent work of a third anthropologist, Mauricio Uribe, from Universidad de Chile in Santiago (Chile), the significance of this ceramic became even more apparent. It revealed how exiled priests from Tiwanaku/ Pumapunku embarked on a daring westward journey from Arica along the Pacific coast. It was during this voyage across the ocean that they became the first inhabitants of Easter Island, around AD 800–900.

In Chap. 1, an article published on May 24, 1981, by Walter Sullivan, a respected journalist from the American New York Times, introduced the research I embarked upon with Peruvian ethnologist Francisco Aliaga forty years ago (refer to Fig. 1. 2). Sullivan wrote, "*(…) Probably the most sensational proposal at the meeting was that many of the most impressive ancient monuments of the world were not carved but cast from stone converted into a plastic form by plant extracts, such as oxalic acid, found abundantly in rhubarb leaves. Examples cited included stones forming the pyramids of Egypt, the ancient Beetle-browed statues of Easter Island, and the great stone structures of the high Andes, such as the famous Gate of the Sun built by the ancient Huanka civilization at Tiahuanaco (…).*"

Sullivan, well-versed in tracking scientific advancements, grasped the breadth of our research project, which we presented at the international conference *Archaeometry* in 1981 in the amphitheater of the Brookhaven National Laboratory near New York. I believe he would have been pleased to learn that after forty years of technological research on the science of geopolymers, we have finally unraveled these mysteries. We have successfully elucidated one of the most widely debated archaeological mysteries in the world. Our groundbreaking findings, which reveal the artificial nature of the rocks used in the monuments of Tiwanaku/Pumapunku, were published in the peer-reviewed scientific journals *Materials Letters* and *Ceramics International* between late 2018 and early 2019. These scientific papers garnered immense attention on the internet, with videos detailing our research receiving over one million views, particularly in Latin America.

Thanks to our discoveries, we now understand how the monuments of Puma-punku, including the mesmerizing megalithic red sandstone terraces and the awe-inspiring H-shaped structural elements crafted from andesite, were shaped using advanced artificial rock fabrication techniques closely related to geopolymer science and technology. This newfound knowledge will undoubtedly aid in comprehending how even more impressive structures were built during this era and beyond. Notably, I am referring to the monumental walls of Sacsayhuaman in Cusco, Peru, as well as other extraordinary monuments found across the globe. It is now the responsibility of South American scientists to embrace this baton and uncover the genius of their ancestors.

Tiahuanaco (Tiwanaku), situated on the shores of Lake Titicaca in Bolivia, is a village renowned worldwide for its enigmatic Sun Gate, temple ruins, and pyramid. Archaeologists assert that this site predates the Inca civilization, dating back to approximately AD 600–700. Next to Tiahuanaco lies Pumapunku, where the ruins of a mysterious pyramidal temple were constructed during the same period. In Chap. 8, I explain the remarkable events of November 2017 when our team ventured to the Pumapunku site and collected samples of what was thought to be red sandstone and andesite. These materials were subjected to an unprecedented analysis using a scanning electron microscope in our geopolymer laboratory. The results of our meticulous examinations unveiled their artificial nature, revealing stark disparities when compared to the local geological resources.

Our rigorous analyses led us to a significant discovery. We already knew that the primary component of the enigmatic structures of Tiwanaku/Pumapunku, namely the Sun Gate and the intriguing "H" blocks, is andesite—a volcanic rock. Astonishingly, we detected the presence of biological organic matter based on carbon (C) and nitrogen (N) within this andesite-made geopolymer. This finding is particularly perplexing because biological carbon-based organic matter should not exist in a volcanic rock that forms under high-temperature conditions. Under such circumstances, it would have simply vaporized, rendering its presence in andesite inconceivable.

Consequently, we can confidently conclude that what was considered to be andesite rock was intentionally manufactured rather than naturally occurring. The identification of this organic matter could pave the way for scientists to perform carbon-14 dating, allowing for a precise determination of the monuments' age. This organic element, a geopolymer derived from carboxylic acids extracted from plants, was deliberately incorporated by human hands to serve as a form of cement.

The colossal blocks of "red sandstone" present a separate problem. Sandstone, a sedimentary rock composed of quartz grains held together by a clay, oxide, or carbonate binder, typically originates from various geological sources. However, none of the potential origins described in Chap. 6 align with the stones found at the archaeological site.

During our field studies in November 2017, we examined several quarries, yet none of them could provide the mammoth 10-m-long blocks that comprise these structures. Furthermore, the local stone available in the vicinity is small in size, rendering it unsuitable for the task at hand. Our SEM investigations of the red sandstone at

Pumapunku unveiled a startling revelation—it contains elements such as high sodium (Na) amounts that do not fit with the local geology. This begs the question: from where did this stone originate? How many hundreds or thousands of kilometers away was it sourced? Moreover, by what means were these immense blocks transported? The electron microscope analysis provides a tantalizing clue—the composition of the "sandstone" suggests the possibility of it being artificial, akin to a ferro-sialate geopolymer, manufactured much like cement.

To produce red sandstone, they took friable and weathered rock, such as red sandstone, from nearby mountains. They then crafted a cement-like mixture using clay, precisely the same red clay used by the Tiwanakans for pottery, and sodium carbonate salts sourced from Laguna Cachi in the Altiplano desert to the south. The addition of lime, as detailed in Chap. 8, completed the formation of the red sandstone.

For the creation of grey andesitic artificial stone, they devised an organo-mineral binder based on plant acids and other natural reagents and combined it with unconsolidated volcanic sand sourced from the nearby Cerro Kapia volcano in Peru. This groundbreaking approach allowed them to replicate the properties of andesite with remarkable accuracy. The resulting cement-like substance was poured into molds and left to harden for several months.

Without a deep understanding of geopolymer chemistry, which studies the formation of rocks through geosynthesis, it would be challenging to discern the artificial nature of these stones. However, mastering this chemistry is not an insurmountable task. In fact, it is an extension of the Tiwanakans' knowledge of ceramics, mineral binders, pigments, and, most notably, their remarkable understanding of their environment. The selection of suitable raw materials was crucial in the creation of these extraordinary monuments, a testament to the Tiwanakans' ingenuity.

Our scientific discovery not only sheds light on the local legends presented to us by Peruvian ethnologist Francisco Aliaga four decades ago but also confirms their validity. According to these legends, the stones were said to be made with plant extracts capable of softening the stone—a notion that archaeologists previously dismissed as nonsensical. However, the evidence uncovered by our team of scientists from France and Peru unequivocally supports the oral tradition. The Tiwanakuans possessed the knowledge to manipulate stones, rendering them malleable before hardening them. With this revelation, we can confidently refute the hypotheses of a lost ancient super civilization or extraterrestrial intervention. The Tiwanakuans were intelligent humans who possessed an intimate understanding of their environment and skillfully harnessed the resources bestowed upon them by nature.

Our study has revealed an intriguing technique used in shaping the architectural components of Tiwanaku/Pumapunku. It appears that a wet sand geopolymer molding technique was employed, possibly resulting in a preform that could be molded into desired shapes. While the geopolymer was still in its malleable state, before complete hardening, artisans of that time would have sculpted it using the traditional tools available, such as wood, stone, and obsidian.

Through our research, we have now identified the geological materials utilized in the creation of the artificial geopolymer andesite volcanic blocks at Tiwanaku/Pumapunku. It is a volcanic sand, known as arena sand, which perfectly suits this purpose. Remarkably, it shares the same mineralogical and chemical composition as the andesite blocks found at the Cerro Khapia volcano on Lake Titicaca. Additionally, we have also determined the geological material used in the construction of the massive megaliths weighing between 100 and 180 tons at Pumapunku. It is a weathered form of red sandstone that easily transforms into sandstone sand due to climatic erosion. This red sandstone is associated with Callamarca (Kallamarka), a historical village that is part of the UNESCO World Heritage, as extensively discussed in Chap. 6. The significance of our discovery cannot be overstated!

Transporting these geological materials over vast distances was well within the capabilities of the construction workers at Pumapunku/Tiwanaku. They employed various methods, such as baskets, rafts, and llama caravans, for this purpose. It was a testament to the ingenuity and labor of *Homo sapiens* on a scale that was entirely feasible.

It is worth noting that significant political events around AD 800 or 900 resulted in the pillaging and looting of the temples at Tiwanaku and Pumapunku. Some accounts suggest that certain priests were exiled to the Pacific Ocean coast, specifically to Arica in present-day Chile. From there, it is said that they traveled from America to participate in the construction of Easter Island—an intriguing historical connection indeed.

The culmination of this research has led me to outline a plausible scenario in Chaps. 10–12, specifically focusing on the settlement of Easter Island and the various phases involved in the construction and manufacturing of the statues. In Chap. 8, I introduced the concept of the "geopolymer of the life-giving world," which serves as the foundation for this proposed system. Allow me to present the four phases or periods that I propose:

- The construction of the Vinapu wall by exiles from Tiwanaku, South America, who arrived from the East around AD 800–900.
- The initiation of the erection of the *Chemamülles* statues, which were planted in the flanks of Rano Raraku by the Mapuches of Chile, coming from the East around AD 1050–1150.
- The establishment of the Moai statues on the Ahus by the Polynesians, arriving from the West around AD 1200–1300. However, the production of *Chemamülles* statues continued during this period.
- The termination of the technology utilizing the geopolymer malleable stone (*Chemamülles* and Moais) according to the Tiwanaku method occurred for reasons that will require further investigation and dating around AD 1500–1600. Potential factors include a scarcity of raw materials and chemical compounds, ecological disasters, destruction of Chilean palm trees (a topic widely discussed among scientists), the arrival of other Polynesians who opposed the established order, a paradigm shift, or a shift in ancestral worship. Due to the inability to adhere to traditional statue-making methods, the statues were then carved from the fragile and friable tuff of the volcano. These statues proved to be less durable, as they were prone to breakage during transport and rapid erosion, while the ones made from artificial rock remained intact and sturdy.

The *Chemamülles* statues were crafted using this method for approximately 450 years, while the production of *Moais* spanned around 350 years. However, the specific events that unfolded in the history of Easter Island following the Polynesian arrival are outside the scope of my expertise. Therefore, I will refrain from delving further into these aspects, as they are actively studied and analyzed by archaeologists, ethnologists, and other experts who have extensively explored the island's history, chronologies, beliefs, traditions, and conflicts between clans.

Indeed, genetic studies have presented varying perspectives on the presence of South American genes among the inhabitants of Easter Island. One study conducted by Moreno-Mayar et al. (2014) indicated the occurrence of a small percentage (8%) of South American genes in the island's population. However, another study published in Current Biology in 2017 by Fehren-Schmotz et al. contradicted these findings, asserting the absence of such genetic characteristics (Fehren-Schmitz et al. 2017).

It is important to note that the archaeological samples analyzed in the 2017 studies originated from a site exclusively occupied by Polynesians, namely the Ahu Nau-Nau at Anakana, which served as a burial place for the Miru royal clan. Gill (2016) had previously established that this royal clan practiced endogamy, or consanguinity, and strictly forbade any intermingling with other clans. This may explain the absence of South American influence in that particular context.

However, the presence of a small percentage of South American genes in unrelated Rapanuis, as detected by Moreno-Mayar's team in 2014, does not necessarily imply colonization of the island by Amerindians arriving from the east. In their article, also published in Current Biology, they stated:

> We detected approximately 8% Native American admixture in unrelated Rapanuis on a genome-wide scale, confirming previous results based on the HLA complex. Although our data suggest that modern-day Rapanuis are the result of a pre-Columbian gene flow event between Native Americans and Polynesians, we cannot, with the data at hand, determine how this gene flow happened. Nevertheless, it is interesting to consider the evidence presented in other fields. There are two main scenarios compatible with the data: Native Americans sailing to Rapa Nui or Polynesians sailing to the Americas and back. It is probably fair to say that the latter scenario is more likely.

My book presents an answer that excludes the latter scenario. According to my research, it was the Amerindians, specifically the Tiwanaku Exiles and later the Mapuches, who were the first to sail to Rapanui, Easter Island. However, patience is required, as there are many taboos surrounding this topic, and numerous influential proponents of the theory attribute the initial and exclusive contribution to the Polynesians.

I repeat here, by way of example, what I wrote in Chap. 1: "Indeed, when it comes to the walls, the construction techniques may exhibit similarities or even outright identity with those found in South America. However, what about the statues, the renowned Moai that stand gazing into the infinite? Could there also be a connection to the civilizations of the east in these enigmatic figures? It appears that the Heyerdahl expedition remains a sensitive topic for the island's inhabitants, one that is not entirely open for discussion. My son Ralph confirmed this in one of his emails from the island, mentioning that he had just visited the museum and noticed the strong emphasis on the Polynesian occupation of the island. Ralph could not find a single word about the Vinapu walls or Thor Heyerdahl's hypothesis.

Actually, the position of ethnologist Robert Langton, as quoted in Chap. 10, aligns with the research presented in my book. Langton states, that his research offers no support for the theory that prehistoric Easter Island was settled solely by Polynesians. On the contrary, he provides strong support for Thor Heyerdahl's long-held claim that there were two American Indian periods of prehistoric culture, and later a Polynesian period.

Thor Heyerdahl, the Norwegian ethnologist (1914–2002), embarked on a daring expedition in 1947 to prove his belief that the islands of Polynesia had been inhabited not only by people from the West but also by people from South America. He set sail on a balsa wood raft named Kon-Tiki with six men on board from the port of Callao near Lima, Peru, on April 28, 1947, with the goal of reaching Polynesia. After 101 days at sea, covering a distance of 4300 nautical miles (7960 km), the Kon-Tiki ran aground on the Raroia atoll in Polynesia. Heyerdahl's successful journey demonstrated that people from South America could have indeed reached Polynesia.

I had the opportunity to meet Thor Heyerdahl in Oslo in 1994. At that time, I had just started my research hypothesis linking the civilizations of the Andes and Easter Island. I was invited by Norwegian shipowner Fred Olsen and Norwegian television to participate in a debate with Heyerdahl. Throughout the weekend, Thor and I engaged in discussions about the remarkable achievements of *Homo sapiens* throughout various eras and continents. The conclusion we reached was that *Homo sapiens* had developed two significant inventions: geopolymer cement and the reed boat, both instrumental in the exploration and expansion of territories.

Today, the intelligence and technical ingenuity of humanity are often underestimated or even scorned by some researchers. Nevertheless, the ability of mankind to find intelligent solutions continues to astonish scientists.

The truth behind these ancient civilizations is even more beautiful and fascinating. I cannot help but admire the remarkable achievements of the inhabitants of the Lake Titicaca region, particularly those from Tiwanaku/Pumapunku in present-day Bolivia, situated at an altitude of nearly 4000 m as well as that of the exiles, alone on this Easter Island lost in the middle of the Pacific. Their ability to adapt commands respect.

References

Fehren-Schmitz L, Jarman CL, Harkins KM, Kayser M, Popp BN, Skoglund P (2017) Genetic Ancestry of Rapanui before and after European contact. Curr Biol 27:1–7. https://doi.org/10.1016/j.cub.2017.09.029

Gill GW (2016) Continuous non-metric characteristics of the early Rapanui. In: Stefan V, Gill GW (ed) Chapter 8. Cambridge University Press, pp 131–154

Janusek JW (2008) Ancient Tiwanaku. Cambridge University Press, New York. ISBN: 978-0-521-01662-9

Moreno-Mayar JV, Rasmussen S, Seguin-Orlando A, Rasmussen M, Liang M, Flam T, Lie BA, Duncan GG, Nielsen R, Thorsby E, Willerslev E, Malaspinas AS (2014) Genome-wide ancestry patterns in rapanui suggest pre-european admixture with native Americans. Curr Biol 24:2518–2525. https://doi.org/10.1016/j.cub.2014.09.057

Additional References

For scientific and technical details or the list of bibliographic references, I recommend that you go to the following Internet addresses

My personal website: www.davidovits.info

Geopolymer Institute: in French, www.geopolymer.org/fr/, in English, www.geopolymer.org

My scientific publications on geopolymers can be downloaded:

Geopolymer Institute at: www.geopolymer.org/library/

Journal on Geopolymer Science Applied to Archaeology at: www.geopolymer.org/library/gpsa

Scientists' network: https://www.researchgate.net/profile/Joseph-Davidovits

Our videos on YouTube at: https://www.youtube.com/user/kadamix